Nanomaterials for Advanced Technologies

Jitendra Kumar Katiyar · Vinay Panwar ·
Neha Ahlawat
Editors

Nanomaterials for Advanced Technologies

Springer

Editors
Jitendra Kumar Katiyar
SRM Institute of Science and Technology
Kattankulathur, Chennai, Tamil Nadu, India

Neha Ahlawat
Jaypee Institute of Information Technology
Noida, Uttar Pradesh, India

Vinay Panwar
Department of Mechanical Engineering
Netaji Subhas University of Technology
Dwarka, Delhi, India

ISBN 978-981-19-1383-9 ISBN 978-981-19-1384-6 (eBook)
https://doi.org/10.1007/978-981-19-1384-6

This Springer imprint is published by the registered company Springer Nature Singapore Pte Ltd.
The registered company address is: 152 Beach Road, #21-01/04 Gateway East, Singapore 189721, Singapore

Contents

Nanomaterials and Their Distinguishing Features

Swati Singh, Naveen Kumar Arkoti, Vivek Verma, and Kaushik Pal

Abstract Nanomaterials have evolved as a fascinating class of materials that encompasses a diverse variety of prototypes with at least one dimension in the 1–100 nm range. The small size and sensible design of nanoparticles can result in extremely high surface areas. As a result of this, nanoparticles have improved features such as high reactivity, strength, surface area, sensitivity, and stability. Nanomaterials can be made with mechanical, magnetic, optical, electrical, and catalytic capabilities that are vastly superior to those of their bulk counterparts. Furthermore, the size, shape, synthesis conditions, and appropriate functionalization of nanomaterials may all be precisely controlled to provide the desired qualities. This chapter gives a brief overview of nanomaterials and how they have been used to progress nanotechnology development throughout history. We discuss and establish nanomaterial classification based on dimensions and materials in particular. The chapter emphasizes the unique characteristics of nanomaterials, such as size and surface area, magnetic properties, quantum effect, and so on. This chapter also discusses nanomaterial advancements and applications in a variety of sectors, including energy harvesting and storage, structural, gas sensing, biomedical and health care, and many more. Finally, we conclude by discussing challenges and future avenues relating to nanomaterials.

Keywords Nanomaterials · Properties · Optical · Surface area · Quantum confinement

1 Introduction

Since the beginning of time, materials have piqued the attention of humans. Rocks were discovered to be capable of breaking objects that were extremely hard to shatter

S. Singh · N. K. Arkoti · K. Pal (✉)
Center of Nanotechnology, Indian Institute of Technology Roorkee, Uttarakhand 247667, India
e-mail: aushik@me.iitr.ac.in; pl_kshk@yahoo.co.in

V. Verma · K. Pal
Department of Mechanical and Industrial Engineering, Indian Institute of Technology Roorkee, Uttarakhand 247667, India

J. K. Katiyar et al. (eds.), *Nanomaterials for Advanced Technologies*,
https://doi.org/10.1007/978-981-19-1384-6_1

with human hands about a million years ago. Stones were the earliest tools, and they have been used to crush and grind food, as well as mortars and pestles, in kitchens and laboratories today. 5000–6000 years ago, it was unintentionally discovered that molten copper could be collected by placing a copper-containing rock over a fire. As a result of this finding, metal ores were reduced to make metals for manufacturing anything from plowshares to swords. New materials for toolmaking were accessible that were harder and lasted longer than stone. The evolution of metals and metallurgy has coincided with our growth and advancement (Murthy et al. 2013).

Materials with exceptional mechanical, physical, and chemical properties are required for new technologies. Materials science and engineering have created materials with diverse qualities by modifying the constituents or microstructure of materials through thermochemical and mechanical processes. As a result, microstructural engineering and structure–property connection studies have become increasingly important. The mechanism whereby the ultrafine microstructures influence the characteristics of materials was being interpreted after the development of theories of lattice distortion and misalignment and the implementation of innovative high-resolution microscopy techniques such as field ion, atomic force, and electron microscopy. These breakthroughs have aided in comprehending the relationship between the properties and structure of solids (Sadik et al. 2014).

Nanotechnology has gained extensive interest over the last century after Richard P. Feynman, the Nobel prize awardee, delivered his lecture at the annual meeting of the American Physical Society on "There's Plenty of Room at the Bottom" in 1959 (Feynman 1992). After this, various discoveries and inventions have been made in nanoscience and technology due to the availability of new tools for the characterization and manipulation of small objects at the atomic scale. Nanotechnology is the controlled manipulation of shape and size at the nanoscale scale (1–100 nm) for the design, characterization, fabrication, and application of structures, devices, and systems having at least one novel/superior trait or characteristic. An essential feature of nanotechnology-produced materials is their higher surface-to-volume ratio, which is desired for many applications. Another essential feature is quantum physics, where nanotechnology enables everyone to create materials with one dimensional (nanowires), two dimensional (nanotubes), or three dimensional (nanoparticles), which are particularly useful in industrial applications (Baig et al. 2021).

In nanotechnology, a nanoparticle (NP) is defined as a tiny constituent array that acts as a single unit and exhibits unique features not seen in bulk materials. The confinement of photons, phonons, and electrons at the nanoscale results in the development of novel biological, physical, and chemical characteristics, making them distinct materials in their very own right. Nanoparticles' (NPs) size, shape, and structure influence their qualities and reactivity. The lowering of dimensionality impacts a range of characteristics, including melting temperature (which is determined by the number of atoms involved in the coordination), conductivity, magnetic properties, optical properties, and reactivity, to name a few. Quantum physics is to blame for these massive shifts. The mean of all quantum forces influencing atomic molecules determines the bulk characteristics of every substance. As particles get progressively smaller, consequently, at a point of time, where averaging becomes ineffective (Yang et al. 2021).

Quantum effects, increased surface area, and self-assembly are among the factors that contribute to nanomaterials' distinctive features. At the nanometer range—especially at the lower end—quantum effects can influence matter's behavior, affecting material's magnetic, electrical, and optical properties (Patra et al. 2011). Second, for the same quantity of material generated in bulk, nanoparticles have a significantly higher surface area. Because the number of active sites increases as particle size decreases, the fraction of surface atoms increases, resulting in an increase in reactivity (Saleh 2020). Finally, self-assembly is a method of constructing an ordered pattern or structure through the organization of components. Supramolecular contacts (ionic, coordination, hydrophobic, hydrogen, and van der Waals bonds) are commonly used in molecular self-assembly, although kinetically labile covalent connections can also be used. When the aggregated and non-aggregated states are equilibrated, the inherent mobility leads to ordered nanostructures with various fascinating qualities, including high sensitivities, self-healing, and error correction to extrinsic stimulant. Because of their unique properties, nanomaterials have applications in multiple fields such as drugs and medication, cosmetics, food packing, biomedical, sensors for environmental stages, electronic and optical, energy harvesting and storage, structural, lubrication, and quantum computing (Foong et al. 2020).

This book gives an in-depth knowledge of how the nanomaterials came into existence and a brief introduction to their unique properties in the first chapter. The second chapter deals with nanomaterials synthesis by top-down and bottom-up methods and characterization techniques like structural, microscopic, composition, optical, magnetic, etc. Chapter 3 deals with different nanomaterials and nanostructures in 0D, 1D, 2D, 3D nanostructures and their simulations of various applications in Chap. 4. The successive chapters discuss the applications of nanomaterials in energy storage and harvesting, structural, sensors and actuators, biomedical, and other discrete areas. The final chapter discusses nanomaterials' future challenges and opportunities in new technologies.

2 Classification of Nanomaterials

Nanomaterials are generally categorized based on their dimension and materials (Fig. 1).

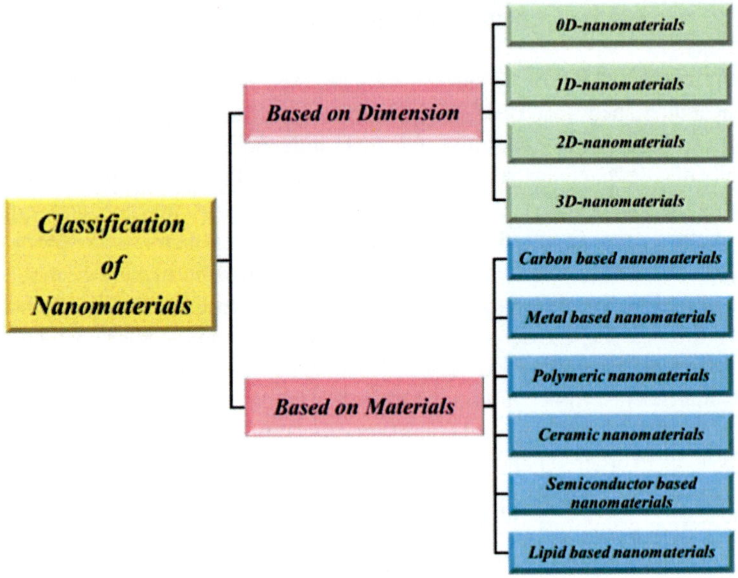

Fig. 1 Classification of nanomaterials

2.1 *Based on Dimension*

2.1.1 Zero Dimensional

Nanomaterials with all of their dimensions in the nanoscale, i.e., scaled below 100 nm, are known as zero-dimensional nanomaterials (0D). There seem to be no dimensions (x, y, z) greater than 100 nm (Saleh 2020). They generally include spherical nanomaterials, hollow sphere, cube, polygon, nanorod, quantum dots (QDs), as well as metal and core–shell nanomaterials.

2.1.2 One Dimensional

One-dimensional nanomaterials (1D) are materials having one non-nanoscale dimension and two nanoscale dimensions (x, y). Needles-shaped nanomaterials are formed as a result of this process. Metals, polymers, ceramics, nanotubes, nanowires, and nanofibers are all examples of 1D nanomaterials.

2.1.3 Two Dimensional

Two-dimensional nanomaterials (2D) contain only one dimension (x) in nanoscale, while the remaining two dimensions are outside the nanoscale. Two-dimensional nanomaterials exhibit plate-like shape. Thin films, nanoplates, and nanocoatings with nanometer thickness are included in this 2D nanomaterials (Pokropivny et al. 2007). They can be single-layered or multilayered, crystalline, or amorphous.

2.1.4 Three Dimensional

Three-dimensional nanomaterials (3D) are materials that not constrained in any way to the nanoscale. Above 100 nm, these materials have three arbitrary dimensions. The bulk (3D) nanomaterials are made up of a variety of nanosize crystals arranged in various orientations (Aversa et al. 2018). They are classically shaped artifacts from the past. They have a length, width, and thickness that are all a few nanometers or more. Multiple nanocrystals are arranged in opposite directions in 3D nanomaterials. Bulk powders, nanoparticle dispersions, nanotubes, fullerenes, nanowire bundles, foams, fibers, polycrystals, honeycombs, and multi-nanolayers are just a few examples of 3D nanomaterials.

2.2 Based on Materials

2.2.1 Carbon-Based Nanomaterials

Carbon-based nanomaterials are primarily hollow spheres, ellipsoids, and tubes made up of carbon. Carbon nanotubes (CNTs) and fullerenes are two prominent groups of carbon-based nanomaterials. Moreover, fullerenes are spherical and ellipsoidal carbon nanomaterials, while carbon nanotubes are cylindrical (Khan et al. 2019). Fullerenes such as allotropic carbon forms are nanomaterials made up of spherical hollow cages. The carbon units in these materials are organized pentagonal and hexagonal, and each carbon is sp^2 hybridized. Electrical conductivity, electron affinity, high strength, structure, and variability all have sparked significant commercial interest.

Carbon nanotubes (CNTs) have a 1–2 nm diameter and are elongated tubular structures. According to their diameter telicity, these are classified as metallic or semiconducting. The structure is similar to that of a graphite sheet rolling on itself. They are referred to as single-walled carbon nanotubes (SWCNTs), double-walled carbon nanotubes (DWCNTs), or multi-walled carbon nanotubes (MWCNTs), depending on the rolled sheets as one, two, or multiple walls (Aqel et al. 2012). Chemical vapor deposition (CVD) technique has recently been used to create them (Elliott et al. 2013). Due to their unique chemical, physical, and mechanical properties, these materials are used to make nanoconjugates for a variety of industrial purposes, including fillers, effective gas adsorbents that work for environmental cleanup, and for different inorganic and organic catalyst's support medium.

2.2.2 Metal-Based Nanomaterials

Metallic precursors are generally used to produce metal-based nanomaterials. Metallic nanomaterials as nanoparticles show localized surface plasmon resonance, a type of resonant electron oscillation that gives them good optoelectrical characteristics. In the visible zone of the electromagnetic spectrum, alkali and noble metal nanoparticles such as gold (Au), copper (Cu), and silver (Ag) have a large absorption band. These nanoparticles with regulated facet, size, and shape are significant in today's cutting-edge materials (Chen et al. 2018). Quantum dots and metal oxides such as copper oxide (CuO) and titanium dioxide (TiO_2) are also included among these nanomaterials. A quantum dot is a nanometer-sized semiconductor crystal that has a core–shell structure in the traditional sense. Quantum dot's optical characteristics are altered by changing their size, shape, and material composition (Hong 2019). Nowadays, metal nanoparticles are used in a wide range of research areas due to their outstanding optical characteristics. Coating of gold nanoparticles is commonly used in FESEM samples to improve the electronic stream, which aids in the acquisition of high-quality FESEM images.

2.2.3 Polymer-Based Nanomaterials

Polymeric nanomaterials are nanoscale solid particles made up of natural or manmade polymers. Dextran, gelatin, pullulan, polylactic acid, chitosan, polylactide-co-glycolide, poly ethylene glycol, and polycaprolactone are some of the polymers commonly employed to make these polymeric nanomaterials as nanocarriers (Yang et al. 2021). These polymers are frequently employed as drug release controls in pharmaceutical and medical applications. Nanocapsules and nanospheres are two main types of polymeric nanoparticles that differ in their shape. Nanocapsules have an oily core in which the chemotherapeutic agent is normally dissolved and have an outer shell of polymer that regulates the drug's release profile from core, while nanospheres are made up of a continuous polymeric network that allows chemotherapeutic agent to be maintained or adsorbed onto their surface (Mansha et al. 2017). These polymeric nanoparticles (PNPs) can be easily functionalized and focused to create new drug delivery systems (DDS) for a variety of ailments.

2.2.4 Ceramic-Based Nanomaterials

Ceramic-based nanomaterials are generally heat-resistant inorganic nonmetallic solids that are made by heating and cooling them repeatedly. These materials have improved structural, optical, superconductive, electrical, ferromagnetic, and ferroelectric characteristics and can be manufactured utilizing physical and chemical processes. They are available in a variety of shapes and sizes, including amorphous, polycrystalline, dense, porous, and hollow. Because of their use in catalysis, photocatalysis, dye photodegradation, and imaging like applications, these nanomaterials are garnering close attention from researchers (Jung et al. 2018).

2.2.5 Semiconductor-Based Nanomaterials

Semiconductor nanomaterials have qualities that are intermediate between metals and nonmetals, allowing them to be used in a variety of applications including solar cells, light-emitting diodes, diodes, transistors, lasers, medical imaging, and quantum computing. The size and shape of these semiconductor NPs or QDs (quantum dots) have a significant impact on their characteristics. They are frequently referred to as "artificial atoms" due to their incredibly small size and the fact that they have unique, discrete, and definite electronic states, as found in naturally occurring atoms/molecules. Photocatalysis, photo-optics, and electrical devices all rely on them. Furthermore, with their acceptable bandgap and band-edge positions, a variety of semiconductor nanomaterials as nanoparticles have been used in water-splitting applications (Singh et al. 2018).

2.2.6 Lipid-Based Nanomaterials

Lipid-based nanomaterials are organic nanoparticles with a lipid unit that can be employed in a wide range of biomedical applications. The diameter of a lipid nanoparticle is 10–1000 nm, and it has a spherical shape. Similarly, to polymeric nanoparticles, lipid nanoparticles have a solid lipid core surrounded by a matrix of soluble lipophilic compounds. To stabilize the nanoparticles outer core, surfactants or emulsifiers were utilized (Rawat et al. 2011). Lipid nanotechnology is a branch of nanotechnology concerned with the design and manufacture of lipid nanoparticles for applications such as delivery of chemotherapeutic agent and release of RNA in cancer therapy.

3 Unique Properties of Nanomaterials

3.1 Size and Surface Area

According to Gleiter's theory, when the size d of a microstructure is reduced to a critical value $d \rightarrow d^*$, the scale length of physical phenomena (phonons and free path length of electrons, etc.; screening length, coherent length, etc.) becomes equal to or compatible with the characteristic size (diameter, thickness, and length) of the microstructure's building blocks.

Generally, a material's properties are defined by a certain "length scale," which is commonly measured in nanometers. If the material's physical size is dropped less than nanoscale, its characteristics alter and become the size and shape sensitive. Nanomaterials will have unique exciting features due to its size effects. The origins of the unique material properties in this size range are not explained by classical physics ideas. Furthermore, because nanocrystals have a large SA (surface area) and a considerable percentage of the atoms in a nanocrystal are on its surface, size effects in chemical and physical characteristics of nanocrystals can arise (30% for a 1 nm crystal, 15% for a 10 nm crystal). As particles and structures become smaller, surfaces and interfaces become increasingly critical (Roduner 2006).

3.2 *Quantum Effect*

Size effects are a unique and exciting feature of nanomaterials. Because nanomaterials are considerably closer to monomolecules and atoms than bulk counterparts, quantum mechanics is required to explain their unneling. In its most basic form, quantum mechanics is a scientific model for explaining the motion and energy of atoms and electrons. The following are the most crucial quantum phenomena, as well as other physical features, that arise in the nanoscale:

Because nanomaterials are so tiny, their mass is shallow; thus, gravitational forces are insignificant. On the other hand, electromagnetic forces play a significant role in influencing how atoms and molecules behave.

Wave-corpuscle duality: The wave-like aspect of the matter is more evident for things of very tiny mass, such as electrons. As a result, electrons behave like waves, and their location is described by a wave (probability) function.

One of the outcomes is a condition known as "unneling." A particle can only enter in via a boundary (potential barrier) provided if there is enough energy to "jump" over it, according to classical physics. If the particle contains less energy than that necessary to leap over the energy barrier, the chances of finding it on the other side of the barrier are nil in classical physics (the "obstacle"). Due to the unneling effect, a particle with less energy than that required to leap the barrier has a finite probability of being positioned on the opposite side of the barrier (Narendra Kumar et al. 2016).

3.3 *Magnetic Properties*

Generally, a material's magnetic behavior is determined by its structure and its temperature. A substance must have a nonzero net spin to experience a magnetic field (transition metals). The size of a traditionally predicted domain is typically approximately 1 m. These materials take on new characteristics at nanometer level. Because of the huge surface-to-volume ratio, a large percentage of molecules have varied magnetic coupling with surrounding molecules, resulting in variable magnetic characteristics (Lu et al. 2007).

Platinum and gold are non-magnetic in bulk; however, they become magnetic at the nanoscale. When Au nanoparticles are coated with the suitable chemical substances, such as thiol, they become ferromagnetic. Magnetic nanoparticles are utilized in imaging, bioprocessing, cooling, and high-density magnetic storage medium, among other uses.

3.4 Thermal Properties

Electrons are the main thermal energy carriers in bulk metals, and nanostructuring can affect their distribution. The free electron density in metallic-type carbon nanotubes, for instance, is low due to spatial confinement, and phonons control their thermal transport behavior. By modifying the accessible energy levels, quantum confinement changes the distribution of carriers. As a result, electrons are now more probable in smaller energy bands that could be used to manipulate thermal properties such as phase transition, melting point (mp), heat capacity, and thermal conductivity (Roduner 2006).

3.5 Mechanical Properties

Planar dislocations in a solid's crystalline structure, such as dislocations, are crucial in defining a material's mechanical characteristics. Because of the supremacy of crystal interfaces and surfaces, it is envisaged that dislocations would play a more diminutive role in explaining nanocrystal qualities compared to the description of microcrystal qualities. A dislocation's free energy is composed of three terms: (1) the free energy deriving from entropy contributions; (2) the elastic strain energy outside the core and extending to the crystal borders; and (3) the core energy (within a radius of around three lattice planes from the dislocation core) (Kumar et al. 2003).

3.6 Electronic and Electrical Properties

The discrete aspect of the energy levels becomes apparent once again when the system size approaches the de Broglie wavelength of the electrons, yet a truly discrete energy spectrum is only found in systems that are confined in all (x, y, z) three dimensions.

Materials that are conducting in nature can become insulators below a threshold length scale when their energy bands stop overlapping. Electrons can tunnel quantum mechanically between two nanostructures that are near together due to their inherent wave-like nature, and if a voltage is applied between two nanostructures that aligns the discrete energy states in the DOS, resonant unneling occurs, dramatically increasing the unneling current.

The quantum confinement effect causes the bandgap to rise as particle's size decreases in the nanoscale regime, causing metal to become a semiconductor as its size decreases. Some nanomaterials' electrical qualities are linked to their distinctive structures and have completely extraordinary electrical properties. Carbon nanotubes, for example, can be either conductors or semiconductors, depending on their nanostructure (Yurkov et al. 2007).

3.7 *Catalytic Properties*

The active surface, where the reaction occurs, is among the most important elements of a catalyst. The active surface increases whenever the catalyst particle size is lowered; the smaller the catalyst particles, the greater the surface-to-volume ratio. Higher the active surface of catalyst, higher the reactivity of surface. It has been appeared that spatial organization of active sites in a catalyst is critical. Both the properties, notably nanoparticle size and molecular structure/distribution, may be controlled via nanotechnology. As a consequence, this method can help the automotive, chemical, petroleum, pharmaceutical, and food industries enhance the design of their catalysts (Bhandari and Knecht 2011).

3.8 *Antimicrobial Activity*

Nanomaterials, particularly metal and metal oxide nanoparticles, offer unique antimicrobial capabilities such as antibacterial, antifungal, and antiviral properties, which make them ideal for application in medical and pharmaceutical devices to combat deadly infections. Furthermore, antimicrobial characteristics of metallic nanoparticles are widely recognized, with many, particularly silver and gold, currently being employed in medical devices to help clean equipment and reduce the transmission of infectious diseases and also in cancer therapy (Bankier et al. 2019). Furthermore, the formation of reactive oxygen species (ROS) by these metallic nanoparticles suppresses the antioxidant defense mechanism and damages the cell membrane, ultimately resulting in cell death (Fig. 2).

4 Applications of Nanomaterials

The presence of unique characteristics of materials at atomic or molecular scale (approximately between 1 and 100 nm) has attracted very much attention and been widely utilized in medicine; targeting drug delivery; artificial implants; sensing

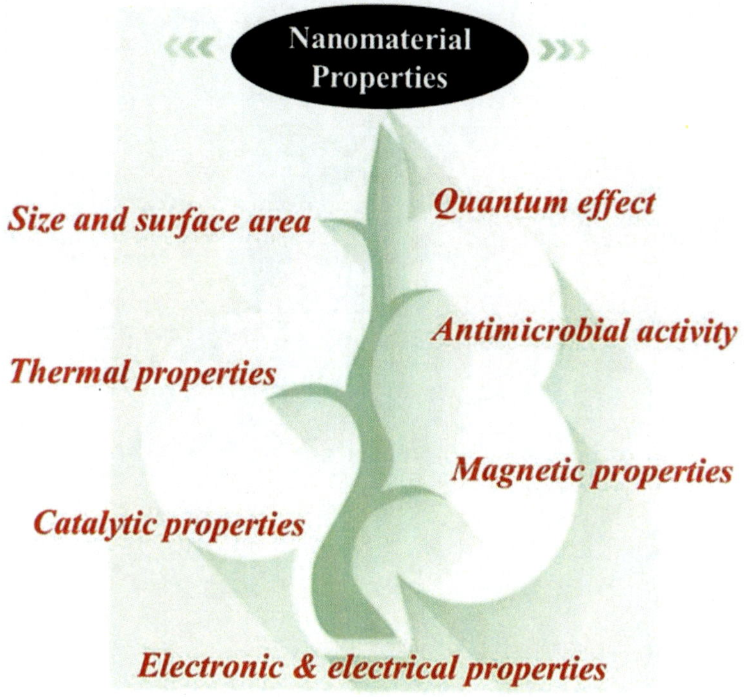

Fig. 2 Properties of nanomaterials

devices for of gas, humidity and biosensors; cancer diagnosis; energy harvesting and storage, environment management, etc. (Barreto et al. 2011); (Yang et al. 2015). With the advancements in technologies, the researchers are able to obtain insights of the chemical, physical characteristics of the nanomaterials, consequently using their potential in various fields. At the one billionth scale of meter, the materials have shown attractive chemical, physical, electrical, and mechanical properties. The nanomaterials usually differ from their bulk forms at this scale. The atoms present at the surface of the materials have different characteristics than the atoms inside the material surrounded with their similar counterparts. When the material is lesser than a critical size, their different characteristics are governed by the principles of quantum physics (Bréchignac et al. 2008).

Nanomaterials are now having multiple applications in every day's life, ranging from cosmetic products to dyes and paints, nutrition, and sports industries (Mehmood 2018). They have their use as antimicrobial and anticancer agents, therapeutic advantages, and nanoparticle-based imaging of central nervous system.

The most predominant applications and technologies based on the use of nanomaterials are categorized as follows: (i) energy storage and harvesting systems (use of

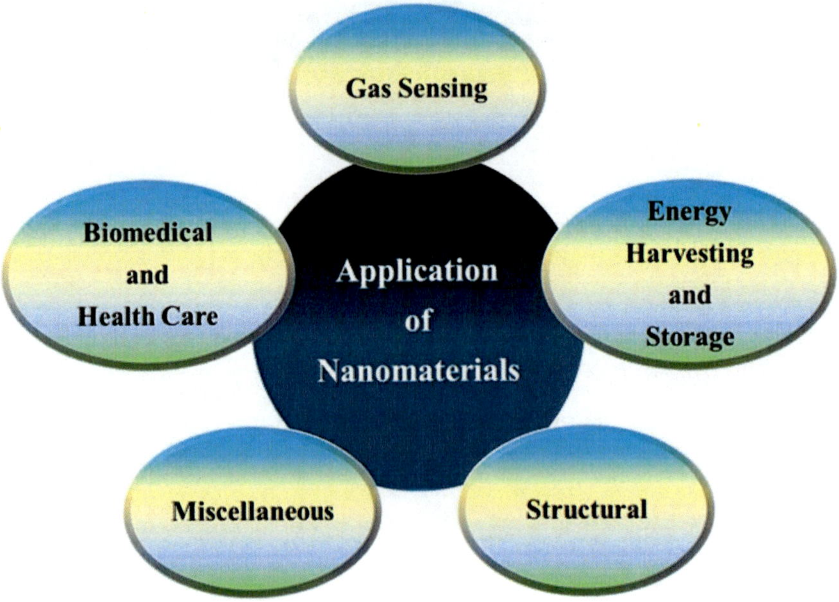

Fig. 3 Applications of nanomaterials in different fields

nanotubes to store hydrogen for different purposes), (ii) nanomaterials and nanocomposites for structural applications (in aerospace and automobile industry), (iii) gas-sensing devices. (iv) biomedical and health care, and (v) other miscellaneous sectors such as environmental management and nanoelectronics (Fig. 3).

The following sections provide an overview of a range of nanomaterials for their effective use in mentioned fields.

4.1 Energy Harvesting and Storage Applications

Energy is the most important part of survival for us human beings on earth. The energy is required for accomplishing the various tasks to meet our needs varying from our food to performing daily activities of our lifestyle. With the increasing demand of energy with the human population, the energy management is need of the hour. Use of fossil fuels as energy sources is one of the major causes of climate change and global warming (Hoel and Kverndokk 1996). The search for green and renewable sources of energy should be a prime focus for sustainable development. The nanomaterials have provided a way to power small electronic products in effective way. The cellulose-based nanostructures (cellulose nanofibrils, i.e., CNFs or nanocrystals), mesoporous structures, thin films, fibers, and their three-dimensional networks are successfully developed for solar cells and photo-electrochemical electrodes (Moon et al. 2011). Cellulose offers its self-assembling feature to attain a wide range of nanoscale configurations suitable for design of energy devices due to its unique structural and chemical characteristics (Wang et al. 2017). Yinhua Zhou et al. developed efficient and recyclable polymer solar cells on optically transparent cellulose nanocrystal substrates for renewable and sustainable energy production (Zhou et al. 2013). The development of nanomaterials has offered an elegant and effective bottom-up approach to take full advantage of tailored chemical composition, structural, electronic, and morphological characteristics in the design of energy devices (Kaur and Pal 2020). It will also enable the production of high efficiency organic photovoltaic materials.

4.2 Structural Applications

With the increased interaction between the nanostructures, having larger surface area employed with ceramic and metal-based materials helps in creating a new generation of lightweight, ultra-high strength, and tough structural nanocomposite materials. The nanostructured tough and hard materials typically used are cobalt/tungsten carbide and iron/titanium carbide-based nanocomposites. Nanoparticles/whiskers/fibers-reinforced polymers are being considered for automobile components. Besides high-strength materials, nanofluids and powders, as well as a wide range of unique morphologies, are being explored. Nanoparticles-based coatings with significantly improved features are being developed. Light structures are the demand of automotive and aerospace applications. Nanostructured metals/alloys are found to exhibit excellent mechanical properties. Nanostructuring of metallic materials helps them achieving superior strength with high ductility (Kumar et al. 2003). The use of severe plastic deformation-based techniques takes the control of nanotwins, grain boundary structures along with the grain size to achieve the concurrence of high strength as well as ductility (Singh and Pal 2021). Nanomaterials have their role in environment control and life support system development space platforms and international space station. The high-performance thermal interface materials with the use of highly conductive carbon nanotubes and metal nanowires are being developed to overcome the thermal management challenges in future aerospace electronics (Qian et al. 2018). The nanoparticle/fiber-reinforced polymer nanocomposites showing high modulus-to-weight and strength-to-weight ratio are in-line of research for their use in aerospace components owing to their multifunctional performance (Panwar and Pal 2020). The carbon-based nanomaterials have been recently investigated for their use as lubricant modifiers and improving the functionality of break pad in automobiles (Ali et al. 2019). The use of nanoparticles, nanofibers, and nanofilms along with the ultrafine-grained nanocrystalline metals has great potential to be used in aerospace and automotive industry. Their employment would help in solving the issue of energy crisis in these sectors.

4.3 Gas-Sensing Applications

The evolution of Internet of Things (IoT) technology having interconnected devices communicating each other raises the need of various sensors. The gas sensors have their use building automation, health care, industrial, and security and public safety-based IoT systems. The use of sensing devices to monitor the presence of gases for environmental control/monitoring, medical diagnosis, etc., is one of the sectors making the use of nanotechnology. The two-dimensional nanomaterials have gained very much attention due to their ultra-high surface-to-volume ratios, excellent semi-conducting performance, thickness dependent chemical and physical properties, and tunable surface activity toward selective gases (Lee et al. 2018); (Rathi et al. 2020). Others having potential in this field are as follows: metal oxide-based nanostructures, metal–organic framework-based nanomaterials, carbons-based nanomaterials, etc.

4.4 Biomedical and Health Care

The increased use of nanomaterials in biomedical application is due to their reduced size in the range of nanometers allowing them their movement into the cells of the living organisms such as human body. The use of nanotechnology and self-assembled nanomaterials in DNA chips and microarrays offers its applications in diagnostics and genetic research. The DNA chips and microarrays-related nanoelectronic devices accompanying thousands of DNA sequences arrayed on a solid substrate are very much useful to various biomedical researchers involved in forensics, drug, and gene discovery studies.

The use of fluorescent polymer-coated nanostructures is very promising in nanobiotechnological research. It has its potential use in development of new biological assays. The nanomaterials offer their targeted treating ability to treat cancer cells without affecting the surrounding cells/tissues. The biocompatible nature of cellulose nanocrystals and carbon-based nanomaterials and the electrospun nanofibers have their potential to be used in tissue engineering and food packaging applications (Pal et al. 2019).

The use of nanomaterials in various health problems is unmatched. Their use as disinfectant and treatment of viral diseases is also prominent. There have been various experimental studies to control severe acute respiratory syndrome coronavirus (Nikaeen et al. 2020). The viral infection is one of the serious concerns because of their ability to mutate and evolve over the time and their fast spreading nature. The viruses have been the cause of many epidemics in the world (Koh and Sng 2010). The use of golds nanoparticles for immunization has been evaluated and led to a significant increment of the peritoneal macrophages respiratory activity in immunized animals (Staroverov et al. 2011). These nanoparticles offer their capability to be engineered for timely and reproducible vaccines. In orthopedics, the use of nanoparticles offers greater control over release of the drug to treat prosthetic joint infections and osteomyelitis and nanotextured implant surfaces and helps to achieve implant osseointegration (Smith et al. 2018).

4.5 *Miscellaneous Applications*

The multidisciplinary field of nanoelectronics offers its use as either single nanostructure (e.g., nanocrystal, nanotube, and quantum dot) for processing optical, chemical, or electrical signals or as assemblies made up of these nanostructures involving their integration for electronic, chemical, biological, optoelectronic, and other applications. It paves a way through for the miniaturization in electronic industry. Nanoparticles-based coatings with highly improved features are being developed for water repellant, antifouling, and flame-retardant applications.

The ultra-small size of nanomaterials is of great interest from the point of view of analytical chemistry and is used at different stages of analytical processes. The CNTs, NPs, and MOFs are used as extraction sorbents, for their good sorption kinetics and for the determination of pollutants in food samples (Socas-Rodríguez et al. 2017).

5 Conclusion

In this chapter, we focused on nanoparticles with major emphasis on their properties, classification, and its applications in various fields. Nanomaterials are classified according to their dimensions and the materials incorporated in their fabrication. Nanomaterials, as a nanoparticle, hold number of desirable properties including optical, catalytic, thermal, magnetic, and electric properties and are between 1 and 100 nm in size. These nanoparticles possess different applications in the field of energy harvesting and storage, structural, biomedical science, gas sensing, and more. It contributes to a healthy environment by delivering cleaner air and water, nanomedicine as well as clean renewable energy for a long-term future. Nowadays, nanotechnology has gotten a lot of attention, and major institutions and organizations are investing more in research and development. Nanotechnology has established itself as a cutting-edge branch of study in which substantial research is being conducted in order to put the technology into practice. It is being tested for a variety of different applications in order to improve the object's efficiency and performance while also lowering the cost so that it is affordable to everyone. This chapter contains a summary of nanomaterials as nanoparticles as well as current information on nanoscience and nanotechnology, which has a bright future because of its efficiency and environmental benefits.

References

I. Ali, A.A. Basheer, A. Kucherova, N. Memetov, T. Pasko, K. Ovchinnikov, V. Pershin, D. Kuznetsov, E. Galunin, V. Grachev, A. Tkachev, Advances in carbon nanomaterials as lubricants modifiers. J. Mol. Liq. **279**, 251–266 (2019). https://doi.org/10.1016/J.MOLLIQ.2019.01.113

A. Aqel, K.M.M.A. El-Nour, R.A.A. Ammar, A. Al-Warthan, Carbon nanotubes, science and technology part (I) structure, synthesis and characterisation. Arab. J. Chem. **5**(1), 1–23 (2012). https://doi.org/10.1016/J.ARABJC.2010.08.022

R. Aversa, M. Hadi Modarres, S. Cozzini, R. Ciancio, A. Chiusole, Data descriptor: the first annotated set of scanning electron microscopy images for nanoscience background & summary. Nat. Publ. Gr. (2018). https://doi.org/10.1038/sdata.2018.172

N. Baig, I. Kammakakam, W. Falath, I. Kammakakam, Nanomaterials: a review of synthesis methods, properties, recent progress, and challenges. Mater. Adv. **2**(6), 1821–1871 (2021). https://doi.org/10.1039/D0MA00807A

C. Bankier, R.K. Matharu, Y.K. Cheong, G.G. Ren, E. Cloutman-Green, L. Ciric, Synergistic antibacterial effects of metallic nanoparticle combinations. Sci. Reports 91 **9**(1), 1–8 (2019). https://doi.org/10.1038/s41598-019-52473-2

J.A. Barreto, W. O'Malley, M. Kubeil, B. Graham, H. Stephan, L. Spiccia, Nanomaterials: applications in cancer imaging and therapy. Adv. Mater. **23**(12), H18–H40 (2011). https://doi.org/10.1002/ADMA.201100140

R. Bhandari, M.R. Knecht, Effects of the material structure on the catalytic activity of peptide-templated Pd nanomaterials. ACS Catal. **1**(2), 89–98 (2011). https://doi.org/10.1021/CS100100K/SUPPL_FILE/CS100100K_SI_001.PDF

Y. Chen, Z. Fan, Z. Zhang, W. Niu, C. Li, N. Yang, B. Chen, H. Zhang, Two-dimensional metal nanomaterials: synthesis, properties, and applications. Chem. Rev. **118**(13), 6409–6455 (2018). https://doi.org/10.1021/ACS.CHEMREV.7B00727

J.A. Elliott, Y. Shibuta, H. Amara, C. Bichara, E.C. Neyts, Atomistic modelling of CVD synthesis of carbon nanotubes and graphene. Nanoscale **5**(15), 6662–6676 (2013). https://doi.org/10.1039/C3NR01925J

R.P. Feynman, There's plenty of room at the bottom. J. Microelectromechanical Syst. **1**(1), 60–66 (1992). https://doi.org/10.1109/84.128057

L.K. Foong, M.M. Foroughi, A.F. Mirhosseini, M. Safaei, S. Jahani, M. Mostafavi, N. Ebrahimpoor, M. Sharifi, R.S. Varma, M. Khatami, Applications of nano-materials in diverse dentistry regimes. RSC Adv. **10**(26), 15430–15460 (2020). https://doi.org/10.1039/D0RA00762E

M. Hoel, S. Kverndokk, Depletion of fossil fuels and the impacts of global warming. Resour. Energy Econ. **18**(2), 115–136 (1996). https://doi.org/10.1016/0928-7655(96)00005-X

N.H. Hong, Introduction to nanomaterials: basic properties, synthesis, and characterization. Nano-sized Multifunct. Mater. Synth. Prop. Appl. 1–19 (2019). https://doi.org/10.1016/B978-0-12-813934-9.00001-3

D.H. Jung, A. Sharma, J.P. Jung, Influence of dual ceramic nanomaterials on the solderability and interfacial reactions between lead-free Sn-Ag-Cu and a Cu conductor. J. Alloys Compd. **743**, 300–313 (2018). https://doi.org/10.1016/J.JALLCOM.2018.02.017

M. Kaur, K. Pal, Potential electrochemical hydrogen storage in nickel and cobalt nanoparticles-induced zirconia-graphene nanocomposite. **31**, 10903–10911 (2020). https://doi.org/10.1007/s10854-020-03641-y

I. Khan, K. Saeed, I. Khan, Nanoparticles: properties, applications and toxicities. Arab. J. Chem. **12**(7), 908–931 (2019). https://doi.org/10.1016/J.ARABJC.2017.05.011

D. Koh, J. Sng, Lessons from the past: perspectives on severe acute respiratory syndrome. Asia-Pacific J. Public Heal **22**(SUPPL. 3), 1863–1952 (2010). https://doi.org/10.1177/1010539510373010

K.S. Kumar, H. Van Swygenhoven, S. Suresh, Mechanical behavior of nanocrystalline metals and alloys. Acta Mater. **51**(19), 5743–5774 (2003). https://doi.org/10.1016/J.ACTAMAT.2003.08.032

E. Lee, Y.S. Yoon, D.J. Kim, Two-dimensional transition metal dichalcogenides and metal oxide hybrids for gas sensing. ACS Sens. **3**(10), 2045–2060 (2018). https://doi.org/10.1021/ACSSENSORS.8B01077

A.H. Lu, E.L. Salabas, F. Schüth, Magnetic nanoparticles: synthesis, protection, functionalization, and application. Angew. Chemie. Int. Ed. **46**(8), 1222–1244 (2007). https://doi.org/10.1002/ANIE.200602866

M. Mansha, I. Khan, N. Ullah, A. Qurashi, Synthesis, characterization and visible-light-driven photoelectrochemical hydrogen evolution reaction of carbazole-containing conjugated polymers. Int. J. Hydrogen Energy **42**(16), 10952–10961 (2017). https://doi.org/10.1016/J.IJHYDENE.2017.02.053

A. Mehmood, Brief overview of the application of silver nanoparticles to improve growth of crop plants. IET Nanobiotechnol. **12**(6), 701–705 (2018). https://doi.org/10.1049/IET-NBT.2017.0273

R.J. Moon, A. Martini, J. Nairn, J. Simonsen, J. Youngblood, Cellulose nanomaterials review: structure, properties and nanocomposites. Chem. Soc. Rev. **40**(7), 3941–3994 (2011). https://doi.org/10.1039/C0CS00108B

G. Nikaeen, S. Abbaszadeh, S. Yousefinejad, Application of nanomaterials in treatment, anti-infection and detection of coronaviruses. Nanomedicine **15**(15), 1501–1512 (2020). https://doi.org/10.2217/NNM-2020-0117/ASSET/IMAGES/LARGE/FIGURE3.JPEG

N. Pal, S. Banerjee, P. Roy, K. Pal, Reduced graphene oxide and PEG-grafted TEMPO-oxidized cellulose nanocrystal reinforced poly-lactic acid nanocomposite film for biomedical application. Mater. Sci. Eng. C **104**, 109956 (2019). https://doi.org/10.1016/J.MSEC.2019.109956

V. Panwar, K. Pal, Influence of addition of selective metallic species on mechanical properties of graphene/acrylonitrile-butadiene-styrene composites. Polym. Compos. **41**(4), 1636–1648 (2020). https://doi.org/10.1002/PC.25485

V. Pokropivny, C VS-MS and E, 2007 undefined Classification of nanostructures by dimensionality and concept of surface forms engineering in nanomaterial science. Elsevier (2007)

M. Qian, D. Yan, J. An, A. Hales, L. Bravo Diaz, M. Waseem Marzook et al., Integrated nanomaterials for extreme thermal management: a perspective for aerospace applications. Nanotechnology **29**(15), 154003 (2018). https://doi.org/10.1088/1361-6528/AAABE1

K. Rathi, K. Pal, K. Rathi, K. Pal, Fabrication of MoSe2–graphene hybrid nanoflakes for toxic gas sensor with tunable sensitivity. Adv. Mater. Interfaces **7**(12), 2000140 (2020). https://doi.org/10.1002/ADMI.202000140

M.K. Rawat, A. Jain, S. Singh, studies on binary lipid matrix based solid lipid nanoparticles of repaglinide: in vitro and in vivo evaluation. J. Pharm. Sci. **100**(6), 2366–2378 (2011). https://doi.org/10.1002/JPS.22435

O.A. Sadik, N. Du, V. Kariuki, V. Okello, V. Bushlyar, Current and emerging technologies for the characterization of nanomaterials. ACS Sustain. Chem. Eng. **2**(7), 1707–1716 (2014). https://doi.org/10.1021/SC500175V

T.A. Saleh, Nanomaterials: classification, properties, and environmental toxicities. Environ. Technol. Innov. **20**, 101067 (2020b). https://doi.org/10.1016/J.ETI.2020.101067

M. Singh, M. Goyal, K. Devlal, Size and shape effects on the band gap of semiconductor compound nanomaterials. **12**(4), 470–475 (2018). 101080/1658365520181473946. https://doi.org/10.1080/16583655.2018.1473946

S. Singh, K. Pal, Investigation on microstructural, mechanical and damping properties of SiC/TiO2, SiC/Li4Ti5O12 reinforced Al matrix. Ceram Int **47**(10), 14809–14820 (2021). https://doi.org/10.1016/J.CERAMINT.2020.10.068

W.R. Smith, P.W. Hudson, B.A. Ponce, S.R. Rajaram Manoharan, Nanotechnology in orthopedics: A clinically oriented review. BMC Musculoskelet Disord **19**(1), 1–10 (2018). https://doi.org/10.1186/S12891-018-1990-1/FIGURES/6

B. Socas-Rodríguez, J. González-Sálamo, J. Hernández-Borges, M.Á. Rodríguez-Delgado, Recent applications of nanomaterials in food safety. TrAC Trends Anal Chem **96**, 172–200 (2017). https://doi.org/10.1016/J.TRAC.2017.07.002

S.A. Staroverov, K.P. Gabalov, L. Dykman, Immunostimulatory Effect of Gold Nanoparticles Conjugated with Transmissible Gastroenteritis Virus. Bull. Exp. Biol. Med. **151**(4), 436 (2011). https://doi.org/10.1007/s10517-011-1350-8

X. Wang, C. Yao, F. Wang, Z. Li, P.X. Wang, C. Yao, F. Wang, Z. Li, Cellulose-based nanomaterials for energy applications. Small **13**(42), 1702240 (2017). https://doi.org/10.1002/SMLL.201702240

G. Yang, C. Zhu, D. Du, J. Zhu, Y. Lin, Graphene-like two-dimensional layered nanomaterials: applications in biosensors and nanomedicine. Nanoscale **7**(34), 14217–14231 (2015). https://doi.org/10.1039/C5NR03398E

K. Yang, S. Zhang, J. He, Z. Nie, Polymers and inorganic nanoparticles: A winning combination towards assembled nanostructures for cancer imaging and therapy. Nano Today **36**, 101046 (2021). https://doi.org/10.1016/j.nantod.2020.101046

G.Y. Yurkov, A.S. Fionov, Y.A. Koksharov, V.V. Koleso, S.P. Gubin, Electrical and magnetic properties of nanomaterials containing iron or cobalt nanoparticles. Inorg. Mater. 438 **43**(8), 834–844 (2007). https://doi.org/10.1134/S0020168507080055

Y. Zhou, C. Fuentes-Hernandez, T.M. Khan, J.C. Liu, J. Hsu, J.W. Shim, A. Dindar, J.P. Youngblood, R.J. Moon, B. Kippelen, Recyclable organic solar cells on cellulose nanocrystal substrates. Sci. Reports 31 **3**(1), 1–5 (2013). https://doi.org/10.1038/srep01536

B.S. Murty, P. Shankar, B. Raj, B.B. Rath, J. Murday, *Textbook of nanoscience and nanotechnology*. Springer Science & Business Media (2013)

M.K. Patra, K. Manzoor, S.R. Vadera, N. Kumar, Functional nanomaterials for sensors and display applications, in *Encyclopedia of nanoscience and nanotechnology*, vol. 14, no. 435, (2011), pp. 385–435

N. Kumar, S. Kumbhat, *Essentials in nanoscience and nanotechnology*, 1st edn. (John Wiley & Sons, Inc, 2016)

E. Roduner, *Nanoscopic materials: size dependent phenomena* (RSC Publishing, Cambridge, UK, 2006b)

C. Bréchignac, P. Houdy, M. Lahmani (eds.), *Nanomaterials and nanochemistry* (Springer Science & Business Media, 2008)

Nanomaterials for Light Harvesting

Sunita Dey and Soumita Talukdar

Abstract All of our day-to-day life needs an energy supply. In an ideal world, we should fulfill all our energy needs from clean, renewable sources and curtain the use of non-renewable, limited stock, greenhouse gas-emitting fuels. Therefore, the present-day energy research is heavily dedicated to mimicking the natural photosynthesis processes in the laboratory by employing synthetic nanocatalyst and modern reaction setups. In this chapter, we discussed the fundamentals behind the artificial light-harvesting processes, the contribution of nanocatalyst in light harvesting and advanced development in this regard. We discussed the standard nanomaterial-dye hybrid model designed for harnessing light to produce H_2 and O_2 by splitting water. Nanomaterials containing semiconductor heterostructures showed excellent water-splitting activity which is improved further in coupling with carbon-based nanomaterials. More challenging, photocatalytic reduction of CO_2 was also exhibited by an array of semiconductors, carbon-based and heterostructures nanomaterials, which are discussed. Photocatalytic degradation of dyes, various organic pollutants, pharmaceuticals, pesticides, etc. by nanorods, organic, inorganic nanoparticles and nanocomposites along with their degradation pathways under various light irradiation is reported. UV light-induced photocatalysis was achieved under visible light irradiation by modifying and tuning the band gap of nanophotocatalysts or using hybrid nanostructures. The impact of various conditions including dispersion phase, reaction flow, temperature, light irradiation and exposed area of the photoreactor setups on the efficiency of photocatalytic activity is discussed briefly. The advancement of hierarchical nanostructure lies in its versatility to be employed in liquid dispersion-based photocatalysis as well as integrated as solar cell devices. Polymer coupled with nanostructures and rare earth-doped nanomaterials also enhance the photovoltaic efficiency of the organic and inorganic nanoparticles. In the end, we presented

Sunita Dey and Soumita Talukdar have contributed equally to this chapter.

S. Dey
Department of Chemistry, University of Cambridge, Lensfield Road, Cambridge CB2 1EW, UK

S. Talukdar (✉)
Department of Chemistry, GD Goenka University, Gurugram 122103, India
e-mail: soumita.talukdar@gmail.com

J. K. Katiyar et al. (eds.), *Nanomaterials for Advanced Technologies*,
https://doi.org/10.1007/978-981-19-1384-6_2

recent developments of semiconductor nanocrystals and hybrid nanocrystals-dye as photorechargeable batteries.

Keywords Nanomaterial · Light harvesting · Photocatalyst · Energy storage · Water splitting

1 Introduction

The ever-expanding industrialization and urbanization are exhausting the fossil fuels stock in unprecedented speed to meet the global energy demands. The massive deforestation and greenhouse gas emission are altering worldwide climate trends which trigger the world to adopt the zero-carbon policy. Burning of the traditional fossil fuels alone is accountable for global CO_2 emission of nearly 9.5 GtC/year (Friedlingstein et al. 2020). Technologies harnessing renewable sources like hydro, wind and solar energy could serve as the alternative to fossil fuel, however, to be least they must excel in energy efficiency and price in comparison to fossil fuels.

Talking about solar energy harnessing, the first thing comes into mind is the natural photosynthesis where plants use only 0.03% of the solar energy and convert it to chemical energy with a quantum yield of near unity (Rao et al. 2016). The obvious research focus is to mimic this natural photosynthesis in the laboratory, and here, we begin this chapter with a discussion on this artificial photosynthesis. Artificial photosynthesis splits H_2O and CO_2 resulting in the production of clean fuel $H_2(g)$ (also $O_2(g)$) and removal of toxic $CO_2(g)$. Development of photocatalyst that can be driven by the energy in the solar visible light is a major part of artificial photosynthesis research, especially involving nanomaterials which can offer tunable size-band gap, short carrier transport length and abundant surface catalytic sites. These photocatalysts could either be suspended in solution (photochemical reaction) or be mounted as photoelectrochemical cell (PEC) to carry out H_2O and CO_2 splitting. This chapter should give readers a detailed description of these methods and importance of nanocatalysts for these purposes. Following this, we discussed a suite of other important photocatalytic applications of nanomaterials such as photodecomposition of toxic pollutants involving reactive oxygen species. The intermittent nature of sun caused an interrupted energy supply; therefore, to realize the full potentials of photocatalytic reactions, development of efficient energy storage device such as battery coupled with photocathode is an ultimate step. We conclude this chapter with the recent findings on photorechargeable batteries and other useful photocatalytic processes.

2 Natural and Artificial Photosynthesis

A fundamental photocatalytic reaction involves absorption of suitable light energy by photocatalyst, generating $e^- - h^+$ pair and catalytic redox reactions with the separated e^- and h^+. Photocatalytic H_2O splitting offers the simple method to produce H_2 by splitting H_2O with four electrons two-step reactions (Eqns. 1 and 2).

$$\text{Oxidation}: 2H_2O + 4h^+ \rightarrow O_2 + 4H^+(1.23\text{V vs. SHE}) \quad (1)$$

$$\text{Reduction}: 4H^+ + 4e^- \rightarrow H_2(0\text{V vs. SHE}) \quad (2)$$

In natural photosynthesis, solar energy is absorbed by two photosystems PSI (P700) and PSII (P680) which consist of chlorophyll-based pigments. PS II and PS I are connected via an electron transfer chain which ensures an efficient charge separation in femtosecond timescales (Fig. 1) (Tachibana et al. 2012; Maitra et al. 2014). Upon shining light, PS II absorbs photon, e^- gets excited and reaches to PSI through the electron transfer chain, and H_2O is oxidized to O_2 and H^+ on PS II by using the h^+ left behind. PS I also absorbs photon, and the excited e^- reduces H^+ and converts NADP to NADPH. This photosystem process is known as Z-scheme, and similar working principle is followed in two-step artificial photosynthesis process (APS) (Fig. 2b).

In APS, photoabsorbers are semiconductor and/or dyes of appropriate band gaps. For rapid electron transfer, additional e^- donor and acceptor catalysts are commonly

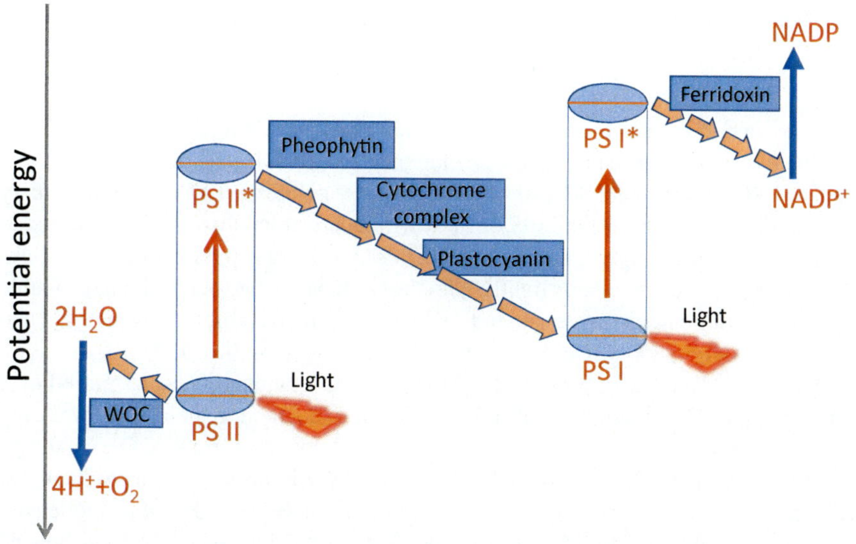

Fig. 1 Schematic showing the natural photosynthesis. WOC refers to Water Oxidation Catalyst and * signify excited state'Adapted from Tachibana et al. 2012 Nature Publishing Group

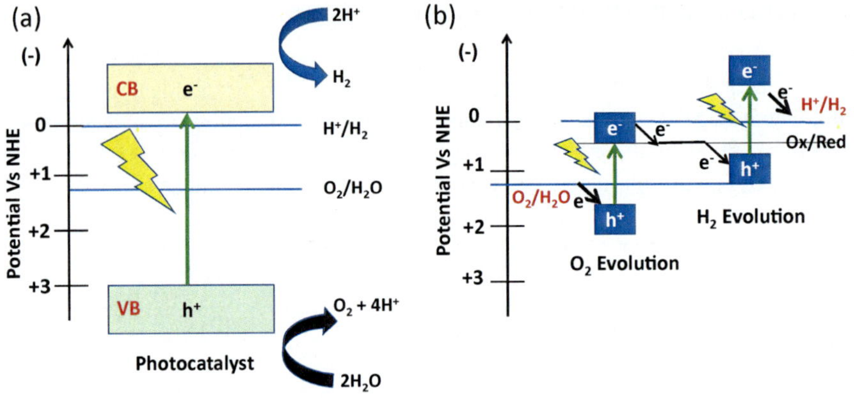

Fig. 2 Schematic showing the **a** one- and **b** two-step artificial photosynthesis process. Adapted from Hisatomi et al. 2014 Nature Publishing Group

adhered on the surface of photoabsorbers. Both photoabsorbers and catalysts should be chosen carefully where the VB (HOMO) and CB (LUMO) positions should satisfy the proton reduction potential (0 V vs SHE), H_2O oxidation potential (1.23 V vs SHE) and rapid e^- transfers between two photoabsorbers. Single-step APS is also well studied which has only one photoabsorber, and both e^- donor and acceptor are adhered on the photoabsorber (Fig. 2).

3 Nanomaterials for Photo(Electro)catalytic Hydrogen Evolution Reaction

For a semiconductor to carry out H_2 evolution reaction (HER) following Eq. 2, the bottom of its conduction band must be more negative than H^+/H_2 reduction potential (0 V vs SHE) (Fig. 2a). Several semiconductors including TiO_2 and ZnO satisfy this condition. However, majority of these semiconductors absorb UV light, and to overcome it, combining dye molecules with this large band gap semiconductor allows the H_2O splitting under visible light. Mallouk et al. have carried out HER under visible light using Ruthenium-dye-sensitized TiO_2 film where dye absorbs visible light and loses e^- to TiO_2 nanoparticles (Youngblood et al. 2009) (Fig. 3a). TiO_2 passes this e^- to Pt counter electrode to produce H_2. Hydrated IrO_x nanoparticles are grafted over this Ru-dye as H_2O oxidation cocatalyst (WOC). The quantum yield (QY) was only ~0.9% attributed mainly to poor charge separation between dye and TiO_2. Coating a thin blocking layer of Nb_2O_5 over TiO_2 shows a slight improvement in photocurrent (Lee et al. 2012). Alternatively, Sun et al. reported dye-sensitized photocathode with organic dye/nanostructured NiO (Lee et al. 2012) where dye loses the photo-induced h^+ to NiO which travels to Pt counter electrode for H_2O oxidation, whereas e^- passes to cobalt-based hydrogen evolving cocatalyst (HEC) for HER (Fig. 3b).

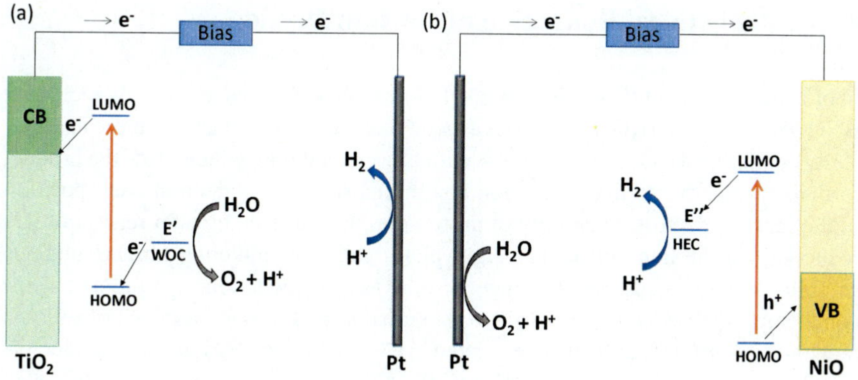

Fig. 3 Energy diagram of a dye-sensitized **a** photoanode and **b** photocathode in a photoelectrochemical water splitting cell. Adapted from (Yu et al. 2015) Royal Society of Chemistry

Similar as dye, quantum dots (QDs) have been used as photosensitizer. Visible light-induced photoelectrochemical HER is observed out of systems like CdTe QDs/ZnO nanowire (Chen et al. 2010), CdSe QDs/N-doped TiO_2 (Hensel et al. 2010), etc. Similarly, carbon QDs are also used as a photosensitizer which ideally be cost-effective and environmental benign in comparison to semiconductor QDs. Chen et al. have used CQDs/graphene oxide (GO) sheet/Pt combination for HER (QY ~ 16%, $\lambda = 420$ nm), where GO sheet acts as fast e- conductor and anchored Pt + CQDs. Pt nanoparticles act as HER cocatalyst (Chen et al. 2016). Carbon nanomaterials such as g-C_3N_4, CNTs also absorb photons (λ ~ 400–500 nm); however, they made superior contribution in photocatalysis as e^- conductors in conjugation with semiconductors and dyes (Singla et al. 2021).

Semiconductors like CdS, MoS_2, $MoSe_2$ are best suited for visible light-induced one-step HER due to their small band gap and ideal conduction band position. However, these materials exhibit weak HER and therefore integrated as PEC device along with dye or other semiconductor and cocatalysts. The 2H-MoS_2 nanoparticles decorated over N-doped graphene sheet exhibit dye-sensitized HER at the rate of 10 mol/g/h. This H_2 productivity increased by >2 times by employing 1 T-MoS_2 (another polytype of MoS_2) which shows metallic conductivity on its own (Maitra et al. 2013). Heterostructure made from a combination of two or multiple semiconductors nanostructures (such as CdS/CdSe/ZnS/TiO_2) could enhance the charge carrier lifetime even by an order of magnitude and offers ample space to manipulate and create the suitable candidates for visible light-induced HER (Chen et al. 2010).

4 Photochemical Reduction of Carbon Dioxide

Photocatalytic reduction of CO_2 (with H_2O) can produce variety of fuels including HCOOH, HCHO, CH_3OH, CH_4, etc. as products. Reduction of CO_2 is more energetically uphill than H_2O reduction and kinetically challenging, which made the product formation selective to many reaction conditions such as conduction band position of the semiconductors, feasibility of proton-coupled electron transfer reaction. CO_2 reduction can be done in liquid or gas phase catalysis, although limited aqueous solubility of CO_2 is another challenge. Similar as H_2O reduction, a plethora of semiconductors (such as TiO_2, ZnO, CdS), quantum dots and carbon-based nanomaterials (e.g., g-C_3N_4, RGO) has been investigated in CO_2 reduction. Various semiconductor-based heterostructures (like ZnO/Cu/CdS, CdSe/Pt/TiO_2) have shown two-step reduction of CO_2 (Kumar et al. 2012; Wu et al. 2017).

5 Other Photocatalytic Processes

As stated above, an absorption of suitable light energy generates $e^+–h^-$ pair, and interestingly, this could induce the reactive oxygen species (ROS) formation, responsible for various photocatalytic reactions (Ma et al. 2019; Ren et al. 2020b). The general reaction scheme of ROS generation and photocatalytic degradation of trypan blue dye are shown in Fig. 4 (Talukdar and Dutta 2016).

Photocatalysis by nanostructures has attracted significant attention in H_2O disinfection (Wang et al. 2017) involving photocatalytic degradation of dyes, toxic pollutants such as phenols, pharmaceutical pollutants, pesticides (Mansor et al. 2020) disposed by textile, leather, tanning and other industries which cause threats to human health and aquatic life. Harvesting sustainable energy for such environmental remediation is of utmost importance. Ma et al. harvested light energy to decompose harmful

Fig. 4 ROS generation in Se-doped ZnS nanoparticle. Reproduced from Talukdar and Dutta (2016)

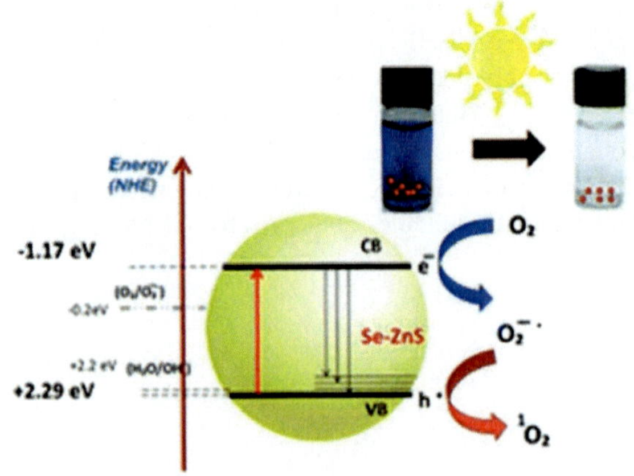

organic dye AO7, using ZnO nanorods with a rate constant of 0.00388 min^{-1} (Ma et al. 2019). Djoko et al. synthesized g-C_3N_4-modified carbon nanomaterials (CNMs) from low cost and easily available tea leaves and urea precursors which act as an efficient UV light-induced photocatalyst for degradation of methylene blue dye with apparent rate constant of 4.7 min^{-1}(Djoko et al. 2020). Various structured nanomaterials have also been designed for efficient photocatalytic H_2O remediation. TiO_2 nanoparticles are most widely used photocatalysts for the inactivation of wide range of microorganisms, such as hepatitis B virus, poliovirus 1, MS2 bacteriophage and herpes simplex virus (Wang et al. 2017). CeO_2-TiO_2 nanocomposite was reported to be used as efficient photocatalyst for degradation of *E. coli* 1337-H under UV and visible light (Wang et al. 2017). Inorganic–organic hybrid of CdS/g-C_3N_4 nanocomposite exhibited complete deactivation of E. coli under solar light irradiation, within 67 min (Baig et al. 2020). Photodegradation pathway of *E. coli* is shown in Fig. 5.

Pharmaceutical drugs are another major toxic H_2O pollutants. Ag- and Fe-modified ZnO nanomaterials exhibited efficient photodegradation of ofloxacina and ciprofloxacin, respectively, under solar light irradiation as shown in Fig. 6 (Majumder et al. 2020), whereas hierarchical structured nanocomposite $Cu_2O/Cu_2(PO_4)(OH)$ exhibited much enhanced, almost 98% photocatalytic degradation of ciproflaxin under solar light irradiation (Beshkar et al. 2021). An array of pharmaceutical and pesticides compounds was reported to be degraded by various metal-doped TiO_2 nanomaterials, under both UV and visible irradiation (Varma et al. 2020).

Wide band gap TiO_2 nanoparticles and CNT + TiO_2 nanomaterials composite exhibit photodegradation of sulfolane, a toxic contaminant used in oil refining and

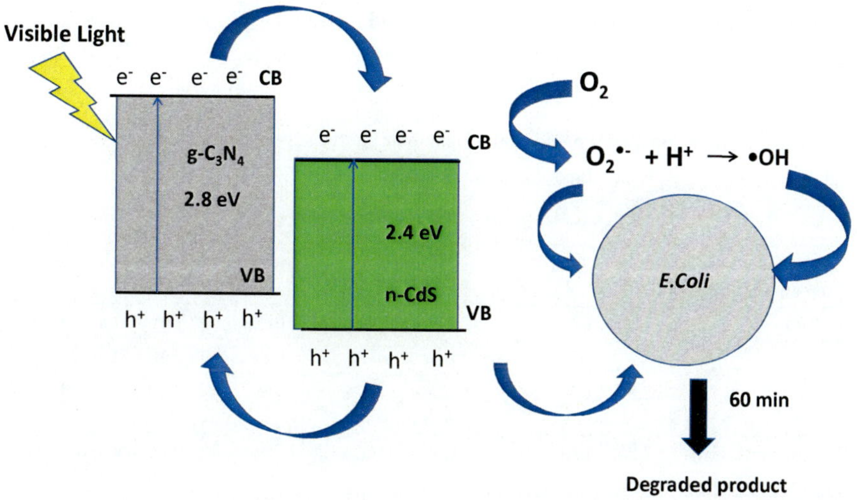

Fig. 5 Visible light-induced photocatalytic degradation of *E. coli* by *n*-CdS@g-C_3N_4 nanocomposite as photocatalyst. Adapted from Baig et al. (2020) Journal of Photochemistry & Photobiology, B: Biology

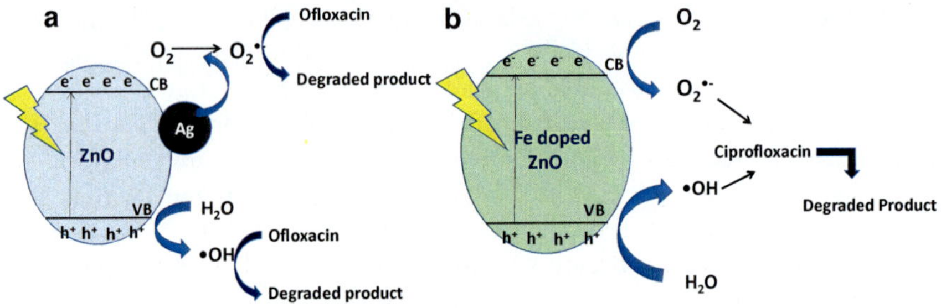

Fig. 6 **a** Degradation pathway of ofloxacin by Ag-modified ZnO NMs under solar light and **b** Degradation pathway of ciproflaxin by Fe-doped ZnO NMs. Adapted from (Majumder et al. 2020) from Environmental Nanotechnology, Monitoring & Management

gas processing, under UVA irradiation with rate constant of 1.93×10^{-2} min^{-1} (Dharwadkar et al. 2021). Carbofuran is one of the most toxic and most commonly used broad-spectrum insecticides which has good stability in water. LED light-induced complete photocatalytic degradation of carbofuran (89%) was achieved by Fe_3O_4-SnO_2-gC_3N_4 nanocomposite, with a rate constant of 0.015 min^{-1} (Mohanta and Ahmaruzzaman 2021), while most of the photodegradation pathways follow pseudo first-order kinetics involving generation of hydroxyl radicals ($OH^{\bullet}$) or superoxide radicals ($O^{2-\bullet}$), which lead to the oxidative decomposition of complex organic molecules. A group of researchers reported, efficient solar light-mediated decomposition of trypan blue dye (complex azo dye) by Se-doped ZnS nanoparticles using singlet oxygen (O^1) as major reactive oxygen species is responsible for the photodegradation, shown in Fig. 4 (Talukdar and Dutta 2016). Various photocatalytic activities by different types of nanomaterials are summarized in Table 1.

Table 1 Various nanomaterial photocatalyst and their photocatalytic degradation activity

Catalyst	Band gap (eV)	Light-induced	Pollutant	Efficiency	References
$Cu_2O/Cu_2(PO_4)(OH)$ heterostructures	2.26	Solar	Ciproflaxin	98%	(Beshkar et al. 2021)
Ag-TiO_2	–	UV	Chloramphenicol	100%	(Varma et al. 2020)
Zr-TiO_2	3.1	UV	Bisphenol A	100%	(Varma et al. 2020)
Fe-TiO_2	2.8	UV	Naproxen sodium	99%	(Varma et al. 2020)
Fe_3O_4-SnO_2-gC_3N_4	2.35	LED	Carbofuran	89%	(Mohanta and Ahmaruzzaman 2021)
Se-ZnS	3.47	Solar	Trypan blue dye	99%	(Talukdar and Dutta 2016)
TiO_2 nanocrystals	3.2	UV	Levofloxacin	90%	(Kansal et al. 2014)

6 Reaction Setup for Various Photocatalytic Process

The design and optimization of the photoreactors are equally important as that of the development of photocatalysts. Experimental parameters play a significant role in the efficacy of the nanomaterial photocatalysts. Therefore, extensive study on the photocatalytic reactions setup needs to be done for the justified comparison of the photocatalytic efficiency. One must also take extreme care about the presence of foreign contaminants in photocatalysis reaction module they could alter activity significantly leading to a misinformation. Metal contaminants enhance the catalytic activity, by reducing the possibility of recombination of charge carriers (Melchionna and Fornasiero 2020). On other hand, an increase in production of CO during the photocatalytic reduction of CO_2 by Cu(I)/TiO_2 photocatalysts is observed on addition of PEG (Li et al. 2010).

In general, photocatalytic reactors could be classified based on (1) photocatalyst deployment and (2) reactor employed.

6.1 *Photocatalyst Deployment*

6.1.1 Dispersed photocatalyst: Dispersed catalytic setup is preferable due to increase in high mass transfer capacity, though there are challenges in isolating the photocatalyst as well as lesser in quantum efficiencies, due to large excess of scattering (Visan et al. 2019).

6.1.2 Immobilized photocatalyst: In this case, low mass transfer can be improved with high flow rate, but it results in turbulence. Reactors with static mixers or can be specially designed for sudden flow changes and reduce the turbulences.

6.2 *Type of Reactor Mode*

6.2.1 Batch Reactors: In case of batch reactors, it is difficult to determine the product composition as the product may get re-adsorbed after their formation and can also get accumulated to initiate side reactions. Therefore, batch reactors are not ideal for large-scale production and for longer duration of the reaction time.

6.2.2 Continuous Flow Reactors: Flow reactors minimize the limitations of product accumulation as that of the batch reactors. It has been experimentally observed that the yield by continuous flow reactors is much higher as compared to the batch reactors (Ali et al. 2019).

Other factors such as temperature, light irradiation, exposed surface area of photocatalyst also significantly affect the photocatalytic activity.

7 Nanomaterials in Solar Cell Applications

The most common photovoltaic devices used for harvesting solar light and its conversion into electrical energy are solar cells. Photoanode is one of the main constituents of any solar cell. Nanomaterials with intrinsic high surface-to-volume ratio are an obvious choice as solar cell material. Recently, 1D nanoparticles have attracted much attention due to their surface architecture and unidirectional pathway for the movement of h^+ and e^- (Bayannavar et al. 2021). CdS nanowires were found to be excellent candidate for dye-sensitized solar cells due to their narrow band gap.

Rare earth-doped fluoride nanomaterials were found to be an excellent candidate for the upconversion in solar cells. The upconversion is the process where emission occurs at higher energy upon excitation of a material with lower energy (Kumar et al. 2020). Therefore, the NIR from the solar radiation can be utilized for upconversion into visible range. Yb, Er, Eu, Tb were reported to be incorporated for application in such solar cells.

Metal nanoparticles such as silver NPs, as plasmon interfaces are currently studied for optimization of light absorption for solar cells (Birant et al. 2020) as it further increases the path length of the light, enhancing light absorption efficiency by 3.5–6.4%.

Carbon sources are abundant in nature; therefore, carbon NMs can also be developed as low cost, efficient components in perovskite solar cells (Moore and Wei 2021). Graphene, GO, reduced GO, rapheme hybrids and CNTs can be extensively used due to their excellent conductivity and photovoltaic properties (Asim et al. 2018; Moore and Wei 2021). These carbon-based materials can be replaced in place of inorganic electrodes due to their excellent conductivity and enhanced charge transport ability.

Organic–inorganic hybrid nanomaterials with suitable organic polymer such as poly(3-hexylthiophene (P3HT) coupled with inorganic accepters (InSb or $MgSnSb_2$ NPs) were also estimated be efficient solar cell candidate. P3HT is extensively used because effective hole mobility is observed in the polymer under solar light irradiation. The inorganic acceptor is chosen in such a way that the band gap is as low as 1.5 eV, and the HOMO lies 0.3 eV lower than the HOMO of P3HT for optimum e^-h^+ mobility (Xiang et al. 2009).

The main efficacy of solar cells depends significantly upon the charge generation, charge separation and their reduced recombination, which can be easily modified by controlling the amount of dopant material, type of ligands and polymers used. Engineered layered structure can also be introduced to reduce the recombination of charge careers incorporating passivating substances or insulators such as metal–insulator–semiconductors junctions (Oener et al. 2016).

8 Photorechargeable Battery

Harvesting solar light to charge electrochemical energy storage devices such as batteries would mitigate the use of any external power supply. A photorechargeable battery is a combination of solar cell and battery which performs two conversion processes: (i) photoenergy to electricity and ii) electricity to electrochemical energy. Upon discharge of a Li-ion battery, Li^+ ions travel from anode to cathode and intercalated on cathode material. During charging, a solar cell material of suitable band gap absorbs light, generates and separates electron–hole pair following which the hole passes to the cathode to push out the intercalated Li ions out of the cathode, and simultaneously, photoelectron transfers through the outer circuit to the anode to neutralize the incoming Li^+ ions (Fig. 7).

The first photorechargeable battery was a three-electrode system composed of $CdSe/S/Ag_2S$ which uses the light-harvesting capability of amorphous CdSe and charge storage chemistry of Ag/S redox pairs (Hodes et al. 1976). Following this, various three-electrode systems were proposed for same purpose, although the main aim remains to convert and store the solar energy in a single device. A two-electrode system is reported by utilizing a hybrid photocathode made of $LiFePO_4$ nanocrystals blended with N719-dye (Paolella et al. 2017). The holes produced due to photoexcitation of N719 dye oxidize $LiFePO_4$ to $FePO_4$ during few hours of light irradiation, while the electron decomposes the battery electrolyte on the surface of Li counter electrode. Despite of its reversible device performance, the photoconversion efficiency of this hybrid photocathode is only 0.06–0.08%. Recently, GeSe nanocrystals have been used as an active material in photorechargeable Li-ion battery, showing an increase in current density of 8 $\mu A/cm^2$ under visible light irradiation (Ren et al. 2020a). Nanostructured, high surface area (photo)electrodes could play an important role in designing photorechargeable batteries by attaining higher light absorption

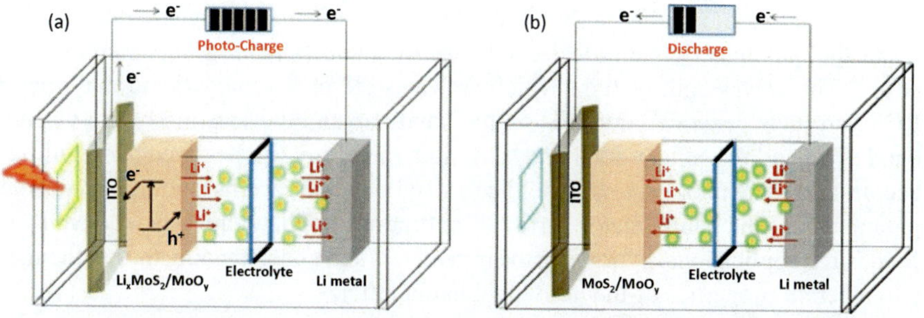

Fig. 7 Schematic showing two-electrode photorechargeable battery consisting of MoS_2/MoO_x nanostructure-based semiconductor as cathode and Li as anode. Fig. b shows that Li^+ ions and electrons are moving toward cathode and battery is producing electricity (discharging) in the absence of light. Fig. (a) shows that cathode is absorbing photon and reverse electrochemical process (charging) is happening. Adapted (Narayanan et al. 2021) American Chemical Society

cum distribution, short carrier transport pathways as well as facile Li^+ ion diffusivity. Narayanan et al. have demonstrated a single nanorod containing MoS_2/MoO_x heterostructure-based solar battery. This two-electrode solar battery utilizes the staggered band alignment of MoS_2/MoO_x which limits the electron-hole recombination and, therefore, pushes the holes toward Li^+ intercalated MoS_2(Narayanan et al. 2021).

In a different approach, polycrystalline metal halide-based 2D perovskite, $(C_6H_9C_2H_4NH_3)_2PbI_4$ has provided both photovoltaic and battery functionalities (Ahmad et al. 2018). However, the system suffered from low photoconversion efficiency of 0.034% (420–650 nm illumination) and Pb–Li alloying issues. To design Pb-free device, photobattery was constructed with 0D $Cs_3Bi_2I_9$ perovskite nanocrystals consisting of thin hexagonal sheets of diameter 10–50 nm (Tewari et al. 2021). Here too, $Cs_3Bi_2I_9$ provides both photoabsorbivity and Li-ion battery activities. In addition, replacing conventional FTO electrode with carbon felt leads to the improvement in photoconversing efficiency to 0.43% under 1 sun for the first discharge. Another recent study has implemented transition metal oxide as a photocathode. The V_2O_5 nanofibers of diameter 20–50 nm are used for efficient charge conduction along the nanofiber length (Boruah et al. 2021). Further mixing of this V_2O_5 with poly(3-hexylthiophene-2,5-diyl) (P3HT) and reduced rapheme oxide (rGO) has carried out efficient charge separation which results in a photorechargeable Li-ion battery of photoconversion efficiency 0.22% under 1 sun. Notably, more benign, cost-effective oxides such as TiO_2 and ZnO have also recently been incorporated as photocathode offering more promises to its future findings (Nguyen et al. 2017; Boruah and De Volder 2021).

9 Conclusion

We can conclude that flexibility of tuning the band structure, absorption of wide spectrum of light, available surface active sights and efficient charge carrier mobility make the nanomaterials an ideal candidate for light-harvesting applications [Ren et al.]. The wavelength of light which activates the photocatalyst, and the type of ROS and fuels generated depends on the band gap and the location of the valence band (VB) and the conduction band (CB) and, therefore, can be tuned and modified accordingly. The photocatalytic yield can also be maximized by optimizing various parameters of the photoreactor. Array of optimum candidate thus can be developed for desired applications, by varying the parameters and understanding the mechanism and inherent properties of the individual nanoparticles.

References

S. Ahmad, C. George, D.J. Beesley et al., Photo-rechargeable organo-halide perovskite batteries. Nano Lett. **18**, 1856–1862 (2018)

S. Ali, M.C. Flores, A. Razzaq et al., Gas phase photocatalytic CO2 reduction, "A brief overview for benchmarking." Catalysts **9**, 727 (2019)

N. Asim, M. Mohammad, M. Badiei, Novel nanomaterials for solar cell devices, in *Nanomaterials for green energy* (Elsevier, 2018), pp. 227–277

U. Baig, A. Hawsawi, M.A. Ansari et al., Synthesis, characterization and evaluation of visible light active cadmium sulfide-graphitic carbon nitride nanocomposite: a prospective solar light harvesting photo-catalyst for the deactivation of waterborne pathogen. J. Photochem. Photobiol. B Biol. **204**, 111783 (2020)

P.K. Bayannavar, A.C. Mendhe, B.R. Sankapal et al., Synthesis of metal free organic dyes: experimental and theoretical approach to sensitize one-dimensional cadmium sulphide nanowires for solar cell application. J. Mol. Liq. **336**, 116862 (2021)

F. Beshkar, M. Salavati-Niasari, O. Amiri, Facile one-pot in situ synthesis and characterization of a Cu2O/Cu2 (PO4)(OH) binary heterojunction nanocomposite for the efficient photocatalytic degradation of ciprofloxacin from aqueous solution under direct sunlight irradiation. Ind. Eng. Chem. Res. **60**, 9578–9591 (2021)

G. Birant, I.M. Ozturk, D. Doganay et al., Plasmonic light-management interfaces by polyol-synthesized silver nanoparticles for industrial scale silicon solar cells. ACS Appl. Nano. Mater. **3**, 12231–12239 (2020)

B.D. Boruah, M. De Volder, Vanadium dioxide-zinc oxide stacked photocathodes for photo-rechargeable zinc-ion batteries. J. Mater. Chem. A (2021)

B.D. Boruah, B. Wen, M. De Volder, Light rechargeable lithium-ion batteries using V2O5 cathodes. Nano Lett. **21**, 3527–3532 (2021)

H.M. Chen, C.K. Chen, Y. Chang et al., Quantum dot monolayer sensitized ZnO nanowire-array photoelectrodes: true efficiency for water splitting. Angew Chemie **122**, 6102–6105 (2010)

L.-C. Chen, T.-F. Yeh, Y.-L. Lee, H. Teng, Incorporating nitrogen-doped graphene oxide dots with graphene oxide sheets for stable and effective hydrogen production through photocatalytic water decomposition. Appl. Catal. A Gen. **521**, 118–124 (2016)

S. Dharwadkar, L. Yu, G. Achari, Enhancement of LED based photocatalytic degradation of sulfolane by integration with oxidants and nanomaterials. Chemosphere **263**, 128124 (2021)

S.Y.T. Djoko, H. Bashiri, E.T. Njoyim et al., Urea and green tea like precursors for the preparation of g-C3N4 based carbon nanomaterials (CNMs) composites as photocatalysts for photodegradation of pollutants under UV light irradiation. J. Photochem. Photobiol. A Chem. **398**, 112596 (2020)

P. Friedlingstein, M. O'sullivan, M.W. Jones et al., Global carbon budget 2020. Earth Syst. Sci. Data **12**, 3269–3340 (2020)

J. Hensel, G. Wang, Y. Li, J.Z. Zhang, Synergistic effect of CdSe quantum dot sensitization and nitrogen doping of TiO2 nanostructures for photoelectrochemical solar hydrogen generation. Nano Lett. **10**, 478–483 (2010)

T. Hisatomi, J. Kubota, K. Domen, Recent advances in semiconductors for photocatalytic and photoelectrochemical water splitting. Chem. Soc. Rev. **43**, 7520–7535 (2014)

G. Hodes, J. Manassen, D. Cahen, Photoelectrochemical energy conversion and storage using polycrystalline chalcogenide electrodes. Nature **261**, 403–404 (1976)

S.K. Kansal, P. Kundu, S. Sood et al., Photocatalytic degradation of the antibiotic levofloxacin using highly crystalline TiO 2 nanoparticles. New J. Chem. **38**, 3220–3226 (2014)

B. Kumar, M. Llorente, J. Froehlich et al., Photochemical and photoelectrochemical reduction of CO2. Annu. Rev. Phys. Chem. **63**, 541–569 (2012)

D. Kumar, S.K. Sharma, S. Verma et al., A short review on rare earth doped NaYF4 upconverted nanomaterials for solar cell applications. Mater. Today Proc. **21**, 1868–1874 (2020)

S.-H.A. Lee, Y. Zhao, E.A. Hernandez-Pagan et al., Electron transfer kinetics in water splitting dye-sensitized solar cells based on core–shell oxide electrodes. Faraday Discuss **155**, 165–176 (2012)

Y. Li, W.-N. Wang, Z. Zhan et al., Photocatalytic reduction of CO2 with H2O on mesoporous silica supported Cu/TiO2 catalysts. Appl. Catal. B Environ. **100**, 386–392 (2010)

J. Ma, J. Ren, Y. Jia et al., High efficiency bi-harvesting light/vibration energy using piezoelectric zinc oxide nanorods for dye decomposition. Nano Energy **62**, 376–383 (2019)
U. Maitra, U. Gupta, M. De et al., Highly effective visible-light-induced H2 generation by single-layer 1T-MoS2 and a nanocomposite of few-layer 2H-MoS2 with heavily nitrogenated graphene. Angew Chemie Int. Ed. **52**, 13057–13061 (2013)
U. Maitra, S.R. Lingampalli, C.N.R. Rao, Artificial photosynthesis and the splitting of water to generate hydrogen. Curr. Sci. 518–527 (2014)
S. Majumder, S. Chatterjee, P. Basnet, J. Mukherjee, ZnO based nanomaterials for photocatalytic degradation of aqueous pharmaceutical waste solutions–a contemporary review. Environ. Nanotechnol. Monit Manag. 100386 (2020)
M.S.A.C. Mansor, M.N.I. Amir, N.M. Julkapli, A. Ma'amor, Gold hybrid nanomaterials: prospective on photocatalytic activities for wastewater treatment application. Mater. Chem. Phys. **241**, 122415 (2020)
M. Melchionna, P. Fornasiero, Updates on the roadmap for photocatalysis. ACS Catal. **10**, 5493–5501 (2020)
D. Mohanta, M. Ahmaruzzaman, Facile fabrication of novel Fe3O4-SnO2-gC3N4 ternary nanocomposites and their photocatalytic properties towards the degradation of carbofuran. Chemosphere **285**, 131395 (2021)
K. Moore, W. Wei, Applications of carbon nanomaterials in perovskite solar cells for solar energy conversion. Nano Mater. Sci. (2021)
N.T. Narayanan, A. Kumar, P. Thakur et al., Photo rechargeable Li-Ion batteries using nanorod heterostructure electrodes
O. Nguyen, E. Courtin, F. Sauvage et al., Shedding light on the light-driven lithium ion de-insertion reaction: towards the design of a photo-rechargeable battery. J. Mater. Chem. A **5**, 5927–5933 (2017)
S.Z. Oener, J. van de Groep, B. Macco et al., Metal–insulator–semiconductor nanowire network solar cells. Nano Lett. **16**, 3689–3695 (2016)
A. Paolella, C. Faure, G. Bertoni et al., Light-assisted delithiation of lithium iron phosphate nanocrystals towards photo-rechargeable lithium ion batteries. Nat. Commun. **8**, 1–10 (2017)
C.N.R. Rao, S.R. Lingampalli, S. Dey, A. Roy, Solar photochemical and thermochemical splitting of water. Philos. Trans. R Soc. A Math. Phys. Eng. Sci. **374**, 20150088 (2016)
C. Ren, Q. Zhou, W. Jiang et al., Investigation of germanium selenide electrodes for the integrated photo-rechargeable battery. Int. J. Energy Res. **44**, 6015–6022 (2020)
X. Ren, D. Philo, Y. Li et al., Recent advances of low-dimensional phosphorus-based nanomaterials for solar-driven photocatalytic reactions. Coord. Chem. Rev. **424**, 213516 (2020b)
S. Singla, S. Sharma, S. Basu et al., Photocatalytic water splitting hydrogen production via environmental benign carbon based nanomaterials. Int. J. Hydrogen Energy (2021)
Y. Tachibana, L. Vayssieres, J.R. Durrant, Artificial photosynthesis for solar water-splitting. Nat. Photonics **6**, 511–518 (2012)
S. Talukdar, R.K. Dutta, A mechanistic approach for superoxide radicals and singlet oxygen mediated enhanced photocatalytic dye degradation by selenium doped ZnS nanoparticles. RSC Adv. **6**, 928–936 (2016)
N. Tewari, S.B. Shivarudraiah, J.E. Halpert, Photorechargeable lead-free perovskite lithium-ion batteries using hexagonal Cs3Bi2I9 nanosheets. Nano Lett. **21**, 5578–5585 (2021)
K.S. Varma, R.J. Tayade, K.J. Shah et al., Photocatalytic degradation of pharmaceutical and pesticide compounds (PPCs) using doped TiO2 nanomaterials: a review. Water-Energy Nexus **3**, 46–61 (2020)
A. Visan, J.R. Van Ommen, M.T. Kreutzer, R.G.H. Lammertink, Photocatalytic reactor design: guidelines for kinetic investigation. Ind. Eng. Chem. Res. **58**, 5349–5357 (2019)
W. Wang, G. Li, D. Xia et al., Photocatalytic nanomaterials for solar-driven bacterial inactivation: recent progress and challenges. Environ. Sci. Nano **4**, 782–799 (2017)
J. Wu, Y. Huang, W. Ye, Y. Li, CO2 reduction: from the electrochemical to photochemical approach. Adv. Sci. **4**, 1700194 (2017)

H. Xiang, S.-H. Wei, X. Gong, Identifying optimal inorganic nanomaterials for hybrid solar cells. J. Phys. Chem. C **113**, 18968–18972 (2009)

W.J. Youngblood, S.-H.A. Lee, Y. Kobayashi et al., Photoassisted overall water splitting in a visible light-absorbing dye-sensitized photoelectrochemical cell. J. Am. Chem. Soc. **131**, 926–927 (2009)

Z. Yu, F. Li, L. Sun, Recent advances in dye-sensitized photoelectrochemical cells for solar hydrogen production based on molecular components. Energy Environ. Sci. **8**, 760–775 (2015)

Transparent Conductive Oxide Nanolayers for Dye-sensitized Solar Cell

Girija Nandan Arka, Shashi Bhushan Prasad, and Subhash Singh

Abstract Dye-sensitized solar cells (DSSCs) stacked in third-generation photovoltaic cells got tremendous limelight for having simple, efficient and economical prospects. The DSSC system comprises of a transparent conductive oxide (TCO) layer coated over a substrate, semiconductive oxide (SCO) film, dye, electrolyte and platinized counter electrode, respectively. The cutting-edge technology involved makes the DSSC distinct from other photovoltaic cell but confronted with a major implication for which the DSSC found underdog photovoltaic cell. DSSC uses light-sensitive dyes for considerable electron generation resulted from photon participation. However, the issue was originated from the sheet resistance when the conventional transparent conductive oxide layer made up of indium tin oxide (ITO) or fluorine tin oxide (FTO) coated on the substrate. This increased sheet resistance potentially discourages the fill factor for which the cell performance potentially deteriorated. Therefore, the conventional transparent conductive oxide layer needs replacement with novel materials to address encouraged fill factor and hence higher power conversion efficiency (PCE). Thus, this article addresses a comprehensive study on transparent conductive oxide layer and addresses its development with insight discussion.

Keywords DSSC · Transparent conductive oxide · Solar cell · ITO · FTO

1 Introduction

To accomplish energy necessasity demanded by the society propelling for discovering the solution through renewable source for which a hive of activity exercised by the researchers. Since renewable energy is the utmost plentiful source of energy, thus, it has remarkable potential to mitigate the energy needs. Solar energy is a great renewable energy source available widely which could be the commendable choice for the necessity. Therefore, solar cells are discovered to translate these lights that

G. N. Arka · S. B. Prasad · S. Singh (✉)
Department of Production & Industrial Engineering, National Institute of Technology Jamshedpur, Jharkhand 831014, India
e-mail: subh802004@gmail.com

J. K. Katiyar et al. (eds.), *Nanomaterials for Advanced Technologies*,
https://doi.org/10.1007/978-981-19-1384-6_3

are available in near-infrared, infrared and visible region, respectively, into electrical energy significantly. Since the available light comprises of 5% ultraviolet ray having wavelength 300–400 nm, 43% visible ray having wavelength 400–700 nm and 52% infrared ray having wavelength 700–2500 nm (Levinson et al. 2005), thus, an idyllic solar cell should entertain all type of wavelength lights for significant photon participation for significant photon conversion. Therefore, many researchers devoted their research for creating a productive and efficient solar cell. Indian government is also investing a lot for the solar cell installation and for the development through research to overpower the conventional consumable fuel. Therefore, it motivates to find the futuristic productive solar cell. In commercial market, silicon-based solar cells are esteemed remarkable and gained significant popularity worldwide. However, silicon-based solar cell is confronted with rigid construction, challenging silicon manufacturing, etc. Therefore, many alternative energy materials and cutting-edge technology have been amalgamated to develop alternative efficient solar cell. Chronological development of solar cell is classified into three generations among which thin film-based solar cell acknowledged a lot for portability, economical, flexible and efficient prospect. Therefore, the current research precisely focused on development of DSSC classified under third-generation solar cells. Another achievement is counted by installing 2000 KWh/year power plant based on DSSC at Switzerland (Fakharuddin et al. 2014). The notion of DSSC was originated by Vogel in 1870 (Balasingam et al. 2013) and further progressed by James in 1887 (Grätzel 2001; Moser 1887), Hishiki and Gerischer in 1965–1968 (Namba and Hishiki 1964; Gerischer et al. 1967) and so on. O'Regan and Grätzel investigation was a phenomenal achievement by reaching energy conversion efficiency 7.1% (O'Regan and Grätze 1991). The DSSC comprises of number of subsystems assembled together to concoct a unit cell and a module. Each subsystem of DSSC is playing a vital role; for DSSC; dynamic robust mechanism acknowledges significant photon ingress, photon participation, photon conversion, respectively. Year by year, momentously the DSSC development thumped a significant interest on each subsystem level to excel its performance potential. Thus, it is important explore in subsystem level to address the associated problem encounter for which the DSSC failed for commercialization.

1.1 Basic Construction of DSSC

The DSSC is an amalgamate of number of subsystems as transparent substrate, transparent conductive oxide layer, blocking layer, semiconducting oxide layer, dye anchoring, electrolyte and transparent conductive oxide-coated platinized counter electrode, respectively. For better visualization, Fig. 1 presents the architecture of DSSC. The transparent substrate is an entrance for solar light to permit into the cell made up of glass or polymeric substrate such as polyethylene terephthalate (PET) and polyethylene naphthalate (PEN). However, research revealed alternative substrates such as metal foil-based, metal mesh-based and carbon composite-based. The transparent conductive oxide layer served twin role and transmits light into

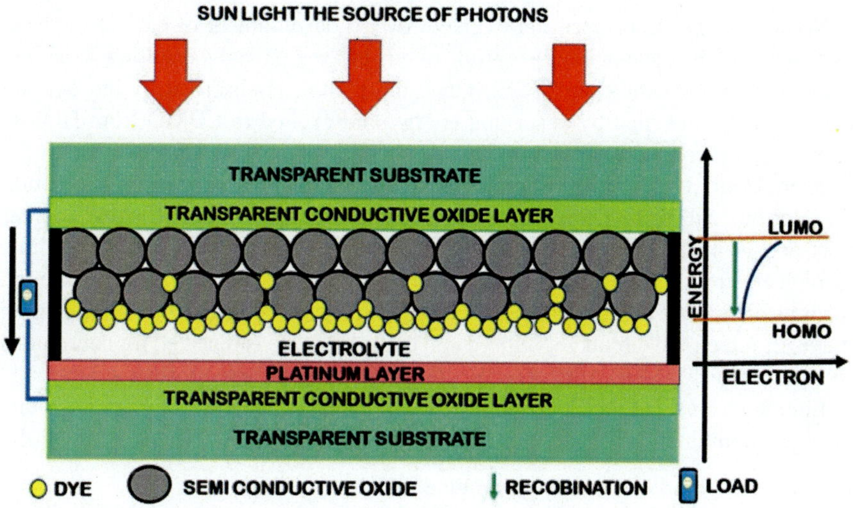

Fig. 1 Basic constructional feature of DSSC showing collaboration of each subsystem for the potential application, whereas the right-side figure denotes optical bandgap of dye from which electron is excited to LUMO and jumps into conduction band of semiconductive layer and is collected at transparent conductive oxide layer for external use

the system and collects electrons from the system and transfers to external load for utilization. The conventional material used for TCOs is fluorine-doped tin oxide (FTO) and indium-doped tin oxide (ITO) (Chen et al. 2010; Fan et al. 2010). The semiconductive oxide layer is used to catch electrons from the least unoccupied molecular orbit (LUMO) of dye and transport rapidly to transparent conducting oxide layer. Frequently used semiconductive oxide layer for DSSC is anatase phase TiO_2 layer owing to its wide optical band gap 3.2 eV (Kazmi et al. 2017). Dye is a light-sensitive material anchored on semiconductive oxide surface which is responsible for electron generation from the highest occupied molecular orbit (HOMO) to LUMO. Electrolyte plays a prime role for electron collection through catalytic activity and recombines electron close to the electrical circuit. Platinum counter electrode is coated over TCO for propelled catalytic activity for recombination at dye. Thus, the study further intensified to explore operational level to find the literature gap of DSSC.

1.2 *Working Mechanism of DSSC*

The DSSC has a light-sensitive dye which would absorb light from 400 to 2500 nm wavelength (visible to infrared region). Prominently, ruthenium-based N719-dye is significantly applied for electron generation due to wide light-absorbing potential from 400 to 920 nm (visible zone) (Shalini et al. 2015).

However, significant attention is given toward development of natural dyes but confronted with inadequate absorption of visible region light resulting degraded power conversion efficiency (PCE). Once the photon is absorbed by the dye, an electron will be excited by absorbing energy and excited to LUMO from HOMO level. Further, conduction band of TiO_2 semiconductive oxide layer will catch the electrons from LUMO and transport to transparent conductive oxide layer. Thereafter, the electrons will be transferred to external circuit for load and collected at counter electrode. From the counter electrode, electrons will be carryforward toward dye hole through electrolyte. The electrons can be transported from counter electrode through catalytic activity by reduction and oxidation reaction. With the electron combination at dye, the electrical circuit gets closed and the same cycle is repeated for continuous operation. However, electron recombination is not desirable from other origin.

Literature revealed that there could be four types of issue which may arise that indulge recombination of electron and hole pair narrated bellow (Haque et al. 2005).

- Electrons from LUMO may recombine with hole at HOMO.
- Electrons at conduction band of semiconductive oxide may recombine with hole at HOMO.
- Electrons at conduction band of semiconductive oxide may recombine with oxidized electrolyte through reduction reaction.
- Electrons at transparent conductive layer may recombine with oxidized electrolyte through reduction reaction.

Therefore, the recombination issue must be anticipated for the efficient DSSC development. Many alterations are reported to address above recombination issue but revealed insignificant result. Thus, transparent conductive oxide layer is playing the major role for electron collection, and it needs to explore in-depth for greater efficiency.

2 Adorable Transparent Conductive Oxide (TCO) Layer

TCO portrays a leading character to collect significant electrons for which it should possess extremely high electrical conductivity. Moreover, it should exceptionally serve to transmit light from visible to infrared region into the system for significant photon participation. Concurrently, TCO film magnanimously supports SCO film mechanically, chemically and electrically by maintaining parity with transparent substrate based on lattice constant and thermal expansion, respectively (Lai et al. 2020). To concoct TCO for the potential application in solar cell, indium tin oxide (ITO) and fluorine tin oxide (FTO) are acknowledged worldwide for having superior optical transmittance and electrical conductivity with optimal sheet resistance (Ye et al. 2014; Bu 2013). Figure 2 is portraying the method to fabricate thin ITO coting on the substrate. ITO gained popularity owing to high transmittance more than 90% with least sheet resistance as low as 10 Ω/sq (glass) which makes it as a potential candidate for TCO. Thus, ITO is the furthermost frequently used transparent conductor, with

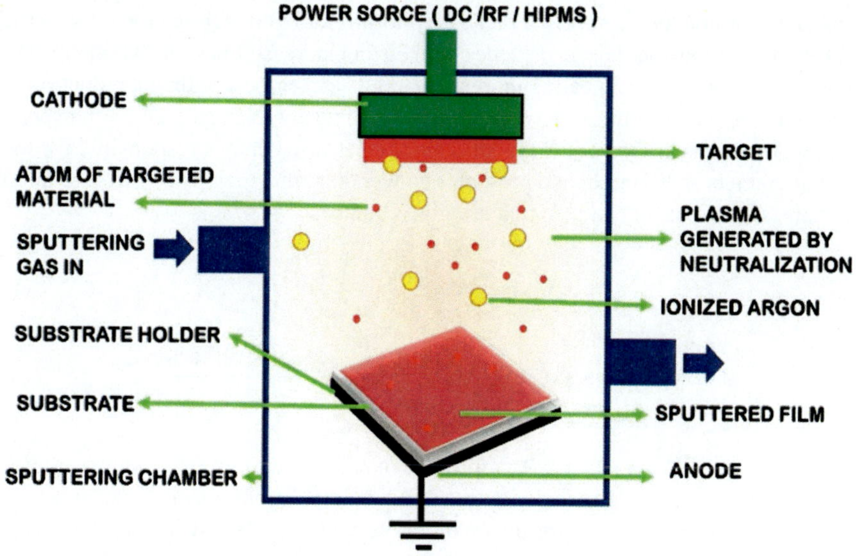

Fig. 2 Semantic diagram of sputtering process to replicate the indispensable components and process to coat a layer over a substrate

sales record of approximately \$1.6 billion in 2013, or 93% of the entire market for transparent conductors (Ghaffarzadeh and Das 2014).

2.1 *Discussion on ITO Technicality*

Primary investigation revealed that ITO is a kind of ceramic which makes the TCO film brittle and leads to crack. Moreover, availability of indium and tin in earth crust is scarce element, and thus, ITO is not sustainable (Rocchetti et al. 2015). Depositing of the ITO film over substrates through vapor phase sputtering process involved a very slow deposition rate approximately 0.01 m/s which makes the cost of film coating per unit length greater. Nevertheless, more than 70% of ITO gets sputtered over the sputtering chamber. Therefore, high conductive ITO is more costly than that of low conductive ITO coating. The typical cost of ITO film coating charged \$5.50/m^2 for 150 Ω/sq whereas \$26.00/m^2 for 10 Ω/sq (Ye et al. 2014).

Thus, radio frequency (RF) or direct current (DC) or magnetron sputtering coating is used to develop thin film coating on the substrates. It incorporates temperature lower than 150 °C for polymeric substrate that consequences a raise in sheet resistance of 15–30 Ω/sq (Weerasinghe et al. 2013). Convincingly, ITO film coating on polymer substrate exhibits sheet resistance 10–16Ω/sq, whereas ITO film coating on nanofiber amorphous silicon unwoven substrate exhibits an elevated sheet resistance of 15–113 Ω/sq (Zobayer et al. 2017). Moreover, a raise of temperature up to 450 °C is needed to anneal the photoanode comprised of TiO_2 nanofilm for DSSC fabrication. One of the report conveys that the sheet resistance could further increase when the ITO-deposited substrate is exposed to annealing more than 300 °C and cause significant reduction to the fill factor and discourage power conversion efficiency (Ngamsinlapasathian et al. 2008). Above discussion conforms that least sheet resistance is responsible for the encouragement of fill factor and promotes conversion efficiency. Theoretical relation is expressed in Eqs. (1) and (2) that are given bellow.

$$\text{fill factor} = \frac{J_M \times V_M}{J_{SC} \times V_{OC}} \tag{1}$$

$$\text{Power conversion efficiency} = \frac{J_{SC} \times V_{OC} \times \text{fill factor}}{P_{in}} \tag{2}$$

where J_M, V_M, J_{SC}, V_{OC} and P_{in} represent maximum current, maximum voltage, short circuit current in mA/cm^2, open circuit voltage in volts and input power in mW/cm^2, respectively, used for the quantitative analysis of DSSC performance.

Moreover, optically ITO behavior not favors for transmitting light from near-infrared region due to its opaque nature (Guo et al. 2016). Thus, the light-harvesting efficiency became reduced for the photon participation.

2.2 Discussion on FTO Technicality

FTO coating is magnanimously employed to construct productive transparent conductive oxide for its wide band gap (~3.6 eV), thermally stable, high transparency and chemically inert, respectively, (Hudaya et al. 2015; Batzill and Diebold 2005; Dauzhenka et al. 2011; Wu et al. 2010) but it has associated problems need to address for the scope of next generation photovoltaic cell. Moving to its technicality, the electrical conductivity of SnO_2 is governed by grain size and type of doping. Doping of fluorine is ideally suitable to retain the crystal structure due to their relatively similar ionic radius. The dopant fluorine atoms substitute the available oxygen sites in the lattice and donate free electrons responsible for creating highly conductive (Wu et al. 2010). Numerous methods that are available to concoct FTO thin film are sputtering, atmosphere pressure chemical vapor deposition, spray pyrolysis deposition and sol-gel, respectively. However, the primary investigation shows that, the electrical conductivity of FTO is inferior to ITO by showing ~10^4 S/cm lesser than ITO (Gassenbauer and Klein 2006). Usually, FTO film coating has thickness range from 500–700 nm frequently incorporated for development of TCO. Nevertheless, for establishing FTO coating, a very high temperature more than 500 °C is needed and for which the cost of the production raised. Moreover, FTO coating over polymer substrate is rather difficult for flexible confinement.

Despite heat resistive, FTO coating on the substrate is showing predominantly high sheet resistance relative to ITO (Lai et al. 2020). Moreover, 60% of DSSC cost is contributed by FTO coating which makes the DSSC expensive (Ayoub et al. 2020). Numerous experiments have been addressed to acknowledge least sheet resistance. To make economical FTO coating, Soe et al. introduced FTO nanopowder synthesized through sol-gel autocombustion route but found insignificance efficiency despite having high fill factor (Soe et al. 2020). Similarly, a novel ultrasonic spray pyrolysis deposition is incorporated by Tuyen et al. to coat FTO over glass, and they produced tetragonal structure at (110), (200) and (211) orientation, respectively, and alter the morphology by polycrystalline with coconut shell-like particles. They are able to clinch as low as 0.7 mΩcm resistivity due to increase in carrier concentration to 1.2 $\times 10^{21}$ cm^{-3} with greater optical transmittance (Tuyen et al. May 2019).

Above comprehensive discussion is directed toward development of highly conductive, transmittance and flexible transparent conductive oxide film with least sheet resistance. Moreover, both FTO and ITO are used to get cracked, while bending test which results reduction in conductivity for flexible DSSC (Jen et al. 2013), hence, should be replaced by alternative encouraging materials. Moreover, expensive cost of the DSSC primarily depends on the transparent conductive oxide film. Hence, alternative low-cost synthesis route should be explored for reduction of overall manufacturing cost of DSSC.

2.3 *Evolution of TCO with Alternative Material*

Remarkable research is endeavored to alter existing conventional TCO with novel materials for the revival of fill factor and for creating least sheet resistance. Recently, aluminum-doped ZnO (AZO) received immense interest in scientific research field for the concocting of productive TCO. The most inspiring properties such as non-toxic, inexpensive, abundant availability and electrical conductivity with optical transmittance unanimously encouraged AZO for the DSSC development.

Doping with Group III elements like In, Ga, Al, etc. into ZnO can considerably lower the electrical resistivity. The resistivity of the ZnO gets reduced due to the partial replacement of the Zn atom by dopant material consequence creation of an additional energy levels. These energy levels are responsible for enriching the optical transmittance. Nevertheless, unlike FTO and ITO, ZnO can withstand high temperature that is required for the removal of organic binder content for making porous interconnected photoanode without significant change in sheet resistance and optical transmittance. However, optimized dopant incorporation is needed for the prolific result.

Hirahara et al. [32] studied the feasibility of AZO by integrating varied Al dopant deposited on quartz glass substrates by RF magnetron sputtering to clinch optimized AZO film for the potential application. They received a notable 2.91% conversion efficiency tested over 0.25 cm^2 under AM1.5 irradiation solar simulator at 4.3 weight% Al-doped ZnO. This result was explored by analyzing bandgap energy and found the band gap increasing with increased dopant concentration. Since doping can be assumed to block the lowest state of conduction band, thus, the band gap is widened by the so-called Burstein–Moss shift. Nevertheless, the prime feature sheet resistance for the AZO film is increased due to ionized and impurity scattering of the electrons by extrinsic dopants. Furthermore, TiO_2 degradation on AZO film is noticed due to imparity of their thermal expansion during calcination required for repelling organic content. Therefore, the mechanical adhesion bond gets weakened and peels of the porous semiconducting oxide layer TiO_2. Nevertheless, the crystallinity of AZO film increases with post-annealing which can be noticed in XRD characterization. It has been noticed that a noticeable c-axis orientation corresponds to vertical growth relative to the substrate after post-annealing 500 °C for 30 min and produced more intense and sharper peak at {002} facet direction. Hence, larger grain size can reduce grain boundary, scattering of light and upsurge carrier lifetime. These collective alterations increase the electrical conductivity as this configuration increases the carrier concentration and Hall mobility and reduces electrical resistivity.

Huang et al. (Huang et al. 2015) introduced AZO TCO over glass and investigate the effect of annealing temperature. They got resistivity and optical transmittance of 1.7×10^{-3} Ωcm and 89%, respectively, for their AZO coating over a glass for post treatment at 500 °C for 30 min. They received esteem Hall mobility with grater carrier concentration and least electrical resistivity which collectively widen the open circuit voltage Voc.

Similar observation is noticed by Chen et al. (2016) that increased carrier concentration and the Hall mobility are collectively empowered to get least resistivity of 1.72×10^{-3} Ωcm resulted from post-annealing at 500 °C which encouraged to clinch increased grain size of the AZO film applied through RF magnetron sputtering onto soda lime glass substrates. Moreover, they found remarkable 88.8% optical transmittance which encourages more light ingress into the system for significant photon participation for greater electron generation through light-sensitive dye anchored over semiconductive oxide layer.

Further, the research direction is moving toward development of AZO by making alteration to enrich greater electrical conductivity with optical transmittance. Qi et al. (Qi et al. 2013) incorporated a novel multilayer of Al-doped ZnO (AZO)/Ag/AZO (AAA) TCO layer over the corning glass by pulsed laser deposition (PLD) technique and direct current (DC)/RF sputtering at room temperature, for the potential application in DSSC. They received remarkable electrical resistivity of2.1×10^{-4} Ωcm and due to which a remarkable 3.25% conversion efficiency was obtained for 400 nm thick AAA film TCO. The result was ascribing to thicker AZO film which potentially prevents undesirable recombination reaction of silver layer with iodine diffusion into AAA layer. Furthermore, the surface roughness of AAA is encouraged to conquer superior adhesion between TiO_2 and AAA film remarkable. Hence, the cell efficiency was promoted by thicker AZO top layer.

Regardless, AZO is confronted with dislocation defects originated from mismatching of lattice constant and thermal expansion with substrates such as glass (Chang et al. Jul. 2013). This technical issue of AZO transparent conductive oxide film leads to deteriorate the cell efficiency drastically. Thus, to revive the cell efficiency by introducing buffer layer between substrate and AZO film could reduce dislocation defect significantly. Further, the buffer layer could be of homogeneous

Table 1 Alternative transparent conductive oxide film investigated for DSSC

Alternative materials	Applied technology	Sheet resistance in Ω/sq	Power conversion efficiency in %	References
Nickel	Hot press	0.18	7.89	(Su et al. 2014)
Mesh-PEDOT: PSS	Dipping method	28.0	2.59	(Berendjchi et al. 2017)
Porous nickel foam		0.00726		(Alami et al. 2020)
AZO	RF sputtering	65	0.25	(Bu 2013)
AZO	RF magnetron sputtering	4.8	2.91	(Hirahara and Nakao 2012)
AZO	RF magnetron sputtering		2.24	(Huang et al. 2015)
AZO	RF magnetron sputtering		3.68	(Chen et al. 2016)
AZO/Ag/AZO	RF sputtering		0.59	(Qi et al. 2013)
IGZO/AZO	RF sputtering	15	3.66	(Lai et al. 2020)
Bi-layer of tin and nickel	Electroless and electroplating		3.44	(Huang et al. 2013)
Silver nanowire embedded	Ultraviolet curing and peeling off	15	0.91	(Fan et al. 2017)
Nickel mesh	Templated electrodeposition and imprint transfer	0.036	5.21	(Khan et al. 2019)
Single-wall carbon nanotube	Dip coating	25	7.0	(Hashmi et al. 2014)

(same crystal structure as that of the AZO) and heterogenous (e.g., Al_2O_3). Ideally, the buffer layer should have excellent optical transmittance within visible region and have exceptional stability under high temperature, respectively, for the potential applications. By considering above indispensable characteristics, indium gallium zinc oxide (IGZO) potentially is acknowledged for its high transparency, exceptional Hall mobility more than 10 cm^2/V. s, wide band gap more than 3.05 eV, amorphous nature which makes it stable under 500 °C, respectively, while selecting heterogenous buffer layer for AZO. Therefore, Lai eta al. (Lai et al. 2020) introduced IGZO heterogenous buffer layer for AZO by RF sputtering. They revealed a notable 62 Ω/sq sheet resistance with 2.6×10^{-3} Ωcm resistivity, respectively, for 424 nm AZO thickness obtained at 130 W sputtering power. Further by introducing IGZO buffer layer by keeping 202 nm buffer layer thickness, they reduce the sheet resistance and resistivity as low as 15Ω/sq and 9.4×10^{-4} Ωcm, respectively, for which they received a notable 3.57% conversion efficiency.

Furthermore, Table 1 is portraying the numerous alterations to replace conventional transparent conductive oxide layer and justifying a promising potential candidate for the DSSC applications.

From above analytical discussion, transparent conductive oxide portrays the most indispensable character to enrich conversion efficiency which ascribes to least sheet resistance. Although many efforts are put forward to replace existing transparent conductive oxide by novel materials, it is found unwelcome due to participation of high processing equipment, dislocation defects, optical reflectance, optical absorption, respectively, (Arka et al. 2021). Since a very few works have been reported to replace existing ITO and FTO transparent conductive oxide layer, hence, it directs a new research scope for the potential applications.

Moreover, the potential of transparent conductive oxide film was fraught with recombination issue by electrolyte penetrated through porous structured semiconductive oxide layer. This problem could be resolved by employing a novel blocking layer over transparent conductive oxide layer to block possible chemical interaction with electrolyte. Ideally, blocking layer should be thin and compact to block electrolyte for the possibility of chemical attack. Numerous materials and technologies are incorporated to address above associated problem for the revival of transparent conductive oxide layer potential. For better visualization, Fig. 3 depicted the importance of blocking layer to avoid possible recombination reaction which eventually reduces the population of electron collected at TCO.

To conquer productive blocking layer, niobium pentoxide titanium dioxide, magnesium oxide, zinc oxide, indium oxide and nickel oxide materials convincingly applied are described by Yu et al., out of which TiO_2 compact layer earned potential respect by integrating SCO with virtuous adhesion and professionally shields

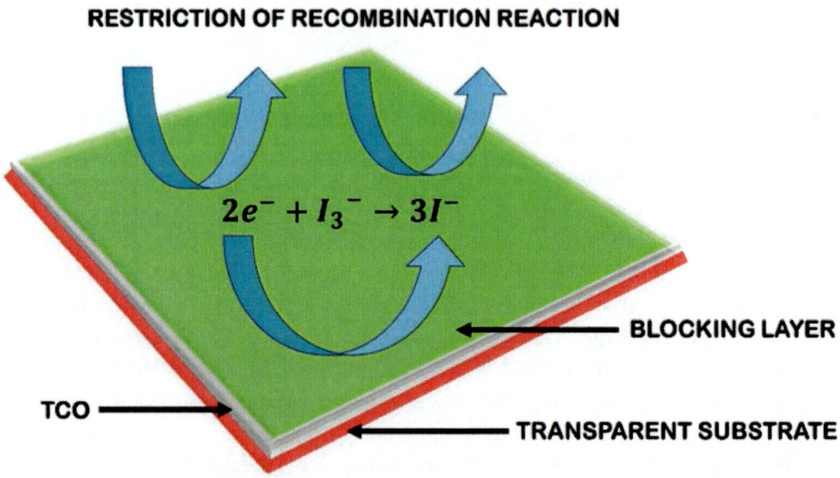

Fig. 3 Portraying the role of blocking layer to protect from redox reaction by iodide and triiodide-based liquid electrolyte penetrated through porous structure of semiconductive oxide layer

the transparent conductive oxide film from the oxidized electrolyte (Yu et al. 2009). The utmost method to produce promising result is by dip coating technique which involves $TiCl_4$ solution treatment at 70 °C for 30 min (Kouhestanian et al. 2020). Moreover, finding of a productive blocking layer is still a paramount challenge.

3 Conclusion

A systematic discussion on dye-sensitized solar cell was exercised to comprehend the significance of transparent conductive oxide layer to collect electrons significantly by avoiding undesired recombination reaction. Mostly, ITO and FTO are conventionally used to make transparent conductive oxide layer for DSSC. However, these conventional materials are questionable for not acquiring significant conversion efficiency due to lower fill factor. Moreover, these materials promote brittleness and increased sheet resistance which further restrict the development of DSSC to clinch high-performance flexible DSSC. Further, the research is directed toward development of transparent conductive oxide layer by alternative promising materials. AZO is found as a promising candidate to replace conventional ITO and FTO. Moreover, AZO demonstrated esteems Hall mobility with grater carrier concentration and least electrical resistivity which is essential for the DSSC potential application. However, AZO confronted with dislocation defects originated from mismatching of lattice constant and thermal expansion with substrates such as glass, thus, could be eliminated by introducing homogenous or heterogenous buffer layer. Indium gallium zinc oxide (IGZO) is potentially acknowledged for the productive buffer layer to address countered mechanical glitches of AZO. Furthermore, many alterations have been reported to replace ITO and FTO by nickel-based, silver nanowire-based, carbon nanotube-based, etc. However, optical reflectance and absorption are still a prime issue to resolve. Hence, this comprehensive discussion adds a new dimension for the research scope to develop a novel productive transparent conductive oxide layer. Numerous 2D materials could be a feasible choice for the development of TCO (Arka et al. 2021; Singh et al. 2021).

Acknowledgements The authors wish to acknowledge the research support provided by Science & Engineering Research Board (SERB), Department of Science and Technology (Grant Number: EEQ/2018/000873) and Ministry of Human Resource and Development (MHRD), Government of India. Also, the authors thank to the Department of Production & Industrial Engineering, NIT Jamshedpur, for the extreme support.

References

R. Levinson, P. Berdahl, H. Akbari, Solar spectral optical properties of pigments—Part II: survey of common colorants. Sol. Energy Mater. Sol. Cells **89**, 351–389 (2005). https://doi.org/10.1016/j.solmat.2004.11.013

A. Fakharuddin, R. Jose, T.M. Brown, F. Fabregat-Santiago, J. Bisquert, A perspective on the production of dye-sensitized solar modules. Energy Environ. Sci. **7**, 3952–3981 (2014). https://doi.org/10.1039/c4ee01724b

S.K. Balasingam, M. Lee, M.G. Kang, Y. Jun, Improvement of dye-sensitized solar cells toward the broader light harvesting of the solar spectrum. Chem. Commun. **49**, 1471 (2013). https://doi.org/10.1039/c2cc37616d

M. Grätzel, Photoelectrochemical cells. Nature **414**(November), 26–32 (2001)

J. Moser, Notiz über die Verstärkung photoelectrischer Ströme durch optische Sensibilisierung. Monatshefte Für Chemie/chemical Mon. **8**(1), 373–373 (1887)

S. Namba, Y. Hishiki, Color sensitization of zinc oxide with cyanine dyes. J. Phyaicnl Chem. **770**(2), 774–779 (1964)

H. Gerischer, M.E. Michel-Beyerle, F. Rebentrost, H. Tributsch, Sensitization of charge injection into semiconductors with large gap. Electrochim. Acta **13**(August 1967), 1509–1515 (1968)

B. O'Regan, M. Grätze, A low-cost, high-efficiency solar cell based on dye-sensitized colloidal TiO2 films. Lett. Nat. **353**, 737–740 (1991). https://doi.org/10.1038/353737a0

H. Chen et al., Plastic dye-sensitized photo-supercapacitor using electrophoretic deposition and compression methods. J. Power Sources **195**(18), 6225–6231 (2010). https://doi.org/10.1016/j.jpowsour.2010.01.009

K. Fan, T. Peng, B. Chai, J. Chen, K. Dai, Fabrication and photoelectrochemical properties of TiO2 films on Ti substrate for flexible dye-sensitized solar cells. Electrochim. Acta **55**, 5239–5244 (2010). https://doi.org/10.1016/j.electacta.2010.04.051

S.A. Kazmi, S. Hameed, A.S. Ahmed, M. Arshad, A. Azam, Electrical and optical properties of graphene-TiO2 nanocomposite and its applications in dye sensitized solar cells (DSSC). J. Alloys Compd. **691**, 659–665 (2017). https://doi.org/10.1016/j.jallcom.2016.08.319

S. Shalini, R. Balasundara, S. Prasanna, T.K. Mallick, S. Senthilarasu, Review on natural dye sensitized solar cells: Operation, materials and methods. Renew. Sustain. Energy Rev. **51**, 1306–1325 (2015). https://doi.org/10.1016/j.rser.2015.07.052

S.A. Haque et al., Charge separation versus recombination in dye-sensitized nanocrystalline solar cells: the minimization of kinetic redundancy. J. Am. Chem. Soc. **127**(10), 3456–3462 (2005). https://doi.org/10.1021/ja0460357

C. Lai, Z. Lee, S. Lin, Y. Chuang, Al-doped ZnO transparent conducting glass with an IGZO buffer layer for dye-sensitized solar cells. IEEE J. Photovoltaics **10**(3), 795–802 (2020). https://doi.org/10.1109/JPHOTOV.2020.2972264

S. Ye, A.R. Rathmell, Z. Chen, I.E. Stewart, B.J. Wiley, Metal nanowire networks: the next generation of transparent conductors. Adv. Mater. **26**(39), 6670–6687 (2014). https://doi.org/10.1002/adma.201402710

I.Y.Y. Bu, Novel fabrication process for flexible dye sensitized solar cell using aluminum doped zinc oxide. Mater. Sci. Semicond. Process. **16**(6), 1730–1735 (2013). https://doi.org/10.1016/j.mssp.2013.07.002

K. Ghaffarzadeh, R. Das, Transparent conductive films (TCF), *Forecasts, Markets, Technologies* (IDTechEx), pp. 2014–2024

L. Rocchetti et al., Cross-current leaching of indium from end-of-life LCD panels. Waste Manag. **42**, 180–187 (2015). https://doi.org/10.1016/j.wasman.2015.04.035

H.C. Weerasinghe, F. Huang, Y.B. Cheng, Fabrication of flexible dye sensitized solar cells on plastic substrates. Nano Energy **2**(2), 174–189 (2013). https://doi.org/10.1016/j.nanoen.2012.10.004

M. Zobayer, B. Mukhlish, Y. Horie, K. Higashi, A. Ichigi, S. Guo, Self-standing conductive ITO-silica nano fiber mats for use in flexible electronics and their application in dye-sensitized solar cells. Ceram. Int. **43**(11), 8146–8152 (2017). https://doi.org/10.1016/j.ceramint.2017.03.140

S. Ngamsinlapasathian, T. Sreethawong, S. Yoshikawa, Enhanced efficiency of dye-sensitized solar cell using double-layered conducting glass. Thin Solid Films **516**(21), 7802–7806 (2008). https://doi.org/10.1016/j.tsf.2008.03.037

W. Guo, Z. Xu, F. Zhang, S. Xie, H. Xu, X.Y. Liu, Recent development of transparent conducting oxide-free flexible thin-film solar cells. Adv. Funct. Mater. **26**(48), 8855–8884 (2016). https://doi.org/10.1002/adfm.201603378

C. Hudaya, B.J. Jeon, J.K. Lee, High thermal performance of SnO2: F thin transparent heaters with scattered metal nanodots. ACS Appl. Mater. Interfaces **7**, 57–61 (2015)

M. Batzill, U. Diebold, The surface and materials science of tin oxide. Prog. Surf. Sci. **79**, 47–154 (2005)

T.A. Dauzhenka, V.K. Ksenevich, I.A. Bashmakov, J. Galibert, Origin of negative magnetoresistance in polycrystalline SnO2 films. Phys. Rev. B **83**, 165309 (2011)

S. Wu, S.L. Yuan, Y. Shi, J. Zhao, J. Fang, Preparation, characterization and electrical properties of fluorine-doped tin dioxide nanocrystals. J. Colloid Interface Sci. **346**, 12–16 (2010)

Y. Gassenbauer, A. Klein, Electronic and chemical properties of tin-doped indium oxide (ITO) surfaces and ITO/ZnPc interfaces studied in-situ by photoelectron spectroscopy. J. Phys. Chem. B **110**, 4793–4801 (2006)

A. Ayoub, N. Mengal, A. Ali, K. Chul, S. Hoon, A rational design of low cost and flexible carbon composite dye sensitized solar cell. Electrochim. Acta **344**, 136050 (2020). https://doi.org/10.1016/j.electacta.2020.136050

O. Soe, P. Kaung, A.A. Thant, Synthesis strategy of F:SnO_2 powder for TCO-less DSSCs fabrication. J. Myanmar Acad. Arts Sci. **XVIII**(2) (2020)

L.T.C. Tuyen, S.-R. Jian, N.T. Tien, P.H. Le, Nanomechanical and material properties of fluorine-doped tin oxide thin films prepared by ultrasonic spray pyrolysis: effects of F-Doping. Materials **12**(10), 1665 (2019)

H. Jen et al., High-performance large-scale flexible dye-sensitized solar cells based on anodic TiO 2 nanotube arrays. ACS Appl. Mater. Interfaces (2013). https://doi.org/10.1021/am402687j

N. Hirahara, M. Nakao, Preparation of Al-doped ZnO thin films as transparent conductive substrate in dye-sensitized solar cell. Thin Solid Films **520**, 2123–2127 (2012). https://doi.org/10.1016/j.tsf.2011.08.100

C.H. Huang, K.S. Chang, C.Y. Hsu, TiO2 compact layers prepared for high performance dye-sensitized solar cells. Electrochim. Acta **170**, 256–262 (2015). https://doi.org/10.1016/j.electacta.2015.04.162

D. Chen, J. Kao, C. Hsu, C. Tsai, The effect of AZO and compact TiO2 films on the performance of dye-sensitized solar cells. J. Electroanal. Chem. **766**, 1–7 (2016). https://doi.org/10.1016/j.jelechem.2016.01.012

J. Qi, Y. Li, T. Duong, H. Choi, S. Yoon, Dye-sensitized solar cell based on AZO/Ag/AZO multilayer transparent conductive oxide film. J. Alloys Compd. **556**, 121–126 (2013). https://doi.org/10.1016/j.jallcom.2012.12.127

S.-J. Chang et al., Triple-junction GaInP/GaAs/Ge solar cells with an AZO transparent electrode andZnOnanowires. IEEE J. Photovolt. **3**(3), 991–996 (2013)

H. Su et al., Highly conductive and low cost Ni-PET flexible substrate for efficient dye-sensitized solar cells. ACS Appl. Mater. Interfaces **6**, 5577–5584 (2014).https://doi.org/10.1021/am406026n

A. Berendjchi, R. Khajavi, A. Akbar, M.E. Yazdanshenas, A facile route for fabricating a dye sensitized solar cell on a polyester fabric substrate. J. Clean. Prod. **149**, 521–527 (2017). https://doi.org/10.1016/j.jclepro.2017.02.075

A.H. Alami, K. Aokal, M. Faraj, Investigating nickel foam as photoanode substrate for potential dye-sensitized solar cells applications. Energy **211**, 118689 (2020). https://doi.org/10.1016/j.energy.2020.118689

Y. Huang, Y. Zhan, S. Cherng, C. Chen, S. Feng, Surface metallization of polyimide as a photoanode substrate for rear-illuminated dye-sensitized solar cells. J. Electrochem. Soc. **160**(9), 581–586 (2013). https://doi.org/10.1149/2.046309jes

R. Fan, C. Zhang, X. Yin, Y. Xiong, S. Xu, Novel flexible photoanode based on Ag nanowire / polymer composite electrode. J. Mater. Sci. Mater. Electron. **28**, 10092–10097 (2017). https://doi.org/10.1007/s10854-017-6770-4

A. Khan et al., Template-electrodeposited and imprint-transferred microscale metal-mesh transparent electrodes for flexible and stretchable electronics. Adv. Eng. Mater. **21**(12), 1900723 (2019). https://doi.org/10.1002/adem.201900723

S.G. Hashmi et al., A durable SWCNT/PET polymer foil based metal free counter electrode for flexible dye-sensitized solar cells. J. Mater. Chem. A **2**, 19609–19615 (2014). https://doi.org/10.1039/C4TA03730H

G.N. Arka, S.B. Prasad, S. Singh, Comprehensive study on dye sensitized solar cell in subsystem level to excel performance potential: a review. Solar Energy **226**, 192–213 (2021). https://doi.org/10.1016/j.solener.2021.08.037

H. Yu, S. Zhang, H. Zhao, G. Will, P. Liu, An efficient and low-cost TiO2 compact layer for performance improvement of dye-sensitized solar cells. Electrochim. Acta **54**, 1319–1324 (2009). https://doi.org/10.1016/j.electacta.2008.09.025

E. Kouhestanian, S. Ahmad, M. Ranjbar, H. Salar, Enhancing the electron transfer process of TiO2-based DSSC using DC magnetron sputtered ZnO as an efficient alternative for blocking layer. Org. Electron. **86**, 105915 (2020). https://doi.org/10.1016/j.orgel.2020.105915

S. Singh, G.N. Arka, S. Gupta, S.B. Prasad, Insights on a new family of 2D material mxene: a review, in *AIP conference proceedings* vol. 2341, no. 1, pp. 040017 (2021). https://doi.org/10.1063/5.0049984

Nanomaterials' Architectural Study in Perovskite Solar Cells

Anshu Varshney and Shreya Sahai

Abstract In this chapter, we premeditate on the architecture and material facet of plasmonic nanoparticles implanted within organic–inorganic halide Perovskite solar cells (PSCs), to yield higher solar absorbance enhancement. The selection of the material of nanoparticle employed within the film is proportional to the enhancement factor of the PSC. Study reveals that the copper nanoparticles produce at par absorbance to other conventional metals such as gold and silver. The utilization of copper could significantly lower this cost without conciliation the solar absorbance of the cell. The location and the size of the particle within the 200-nm-thick perovskite film are also analysed to get better solar absorbance of the designed solar cell. Study also highlights the impact of tailored plasmonic nanoparticles including nanospheres, nanocubes, nanocylinders, nanorods, nanotriangular plates, etc. rooted within the film. This architectural study is extended to the different inclination of a particle within the film, concerning the source of light. Different volumes of these copper nanoparticles are placed in the heart of the film to verify the enhancement of the light trapping efficiency of these designed cells. In the later section of the chapter, two nanoparticles in a unit cell as a dimer are taken into account. The anomalous effects on solar absorbance enhancement with the insertion of dielectric coated plasmonic copper nanoparticle within hybrid organic–inorganic halide perovskite solar cells have also been observed. The metal-dielectric core-shell nanoparticles have been inserted into the active layers of thin-film organic solar cells to improve conversion efficiency.

Keywords Plasmonic nanoparticles · Solar absorbance · Core shell nanoparticle · FDTD · Perovskite solar cells

---

A. Varshney (✉)
Department of Physics and Material Science and Engineering, JIIT, Noida 201307, India
e-mail: anshu.varshney@jiit.ac.in

S. Sahai
Department of Electrical and Computer Engineering, University of Florida, Gainesville, Florida 32608, USA
e-mail: shreya.sahai@ufl.edu

J. K. Katiyar et al. (eds.), *Nanomaterials for Advanced Technologies*,
https://doi.org/10.1007/978-981-19-1384-6_4

1 Introduction

Methylammonium lead halide (MAPbI3) perovskite-based solar cells (PSCs) (Green et al. 2014) have derived a lot of attention in the past few years due to the high transparency, efficiency, direct bandgap property, high power conversion efficiency, and small exciton binding energy of perovskite material. Despite their high performance, a recent analysis of these systems has listed a series of their major drawbacks, including low absorption edge around infrared region and high lead toxicity. Metal nanostructures are suggested to enhance the light matter interaction and as a result the light trapping capability of PSCs increases at longer wavelengths, based on their characteristic property of surface plasmon resonance (SPR).These nanostructures consist of free conduction electrons which oscillate with the frequency of incident radiance. SPR depends upon numerous factors including size, morphology, and dielectric medium in which they are placed. Usually, the resonance wavelength of metals nanostructures observes in the red region of the visible band. Near- and far-field effects produced by these nanostructures lead to intensity enhancement and strong scattering around them, respectively. This phenomenon magnifies the light trapping ability of these cells and thus improving its efficiency. High percentage of lead due to the enhanced thickness of the perovskite layer poses a major threat to its commercialization. Too thin films on the other hand also lead to unoptimized light collection due to minimal interaction of perovskite with the incoming light wave. To optimize this effect without compromising the absorption efficiency of these cells, the thickness of the perovskite layer is set to 200 nm.

To explore the plasmonic effect in perovskite solar cells, several studies have been conducted so far. Core–shells nanoparticles have been also exploited to augment the absorption efficiency of PSCs. The work done by Saliba studies core–shell Au@SiO2 nanoparticles within the perovskite film of PSCs to achieve enhanced photocurrent with an efficiency as high as 11.4% (Zhang et al. 2013). Interestingly, at the optimum concentration of these particles, a minimum increase in the light trapping capacity of the device is found. The author also introduces a novel photon recycling scheme with the incorporation of silver nanomaterials including core–shell Ag@TiO2 nanoparticles, within low-temperature solar cells, achieving an efficiency of 16.3% (Saliba et al. 2015).At the same time, work by Dabirian explores dye-sensitized solar cells with a monolayer of core–shell Silica@Ag nanoparticles to enhance the current density of short circuit of a cell by 38% (Dabirian et al. 2016). Moreover, the work lead by S. Carretero-Palacios theoretically highlights the impact of embedding plasmonic nanoparticles of varying sizes, shapes, and concentration within the perovskite film (Carretero-Palacios et al. 2016).Due to the chemical stability of core–shell particles within different systems, they have become crucial subject of interest among nanotech researchers around the globe. The research lead by N. Pathak analyses the enhancement in absorption efficiency with the incorporation Au@SiO2 nanoparticles as a function of shell thickness, achieving an efficiency as high as 18% (Pathak et al.

2017). Inverted perovskite solar cells containing spectrally tuned gold nanorods indicate a significant increase in the device photocurrent owing to the LSPR excitation at longer wavelengths (Cui et al. 2016).

Few other works have also suggested a combinational use of gold NPs and magnesium oxide passivation layer to improve the stability of the PSCs device under UV light. Au NPs embedded within mesoporous TiO2 layer provide enhanced absorption associated with effective photon management, while MgO layer minimizes photonic and energy losses leading to an efficiency of 16.1% (Luo et al. 2017). Work done by Batmunkh demonstrates enhanced electrostatic potential with the incorporation of Au@SiO2nano stars at the interface of the compact TiO2 electron selective contact and the mesoporous TiO2 film achieving 17.55% efficiency (Batmunkh et al. 2017). Maximum efficiency of 20.1% is achieved by a device incorporated with Au@TiO2 nanorods within TiO2 layer of the cell (Fan et al. 2017). The performance of the cell is improved by 5%.These works highlight that the solar enhancement of PSCs depends upon the numerous parameters. Variation of the size, shape, and concentration of plasmonic nanoparticles impacts the absorption efficiency.

As stated previously, it was suggested that gold nanoparticles or silver nanoparticles for solar absorbance enhanced the absorbance. The incorporation of gold or silver nanoparticles drastically increases the production costs of PSCs. We work out on the cheaper yet effective alternative of these expensive materials. Silver and gold can be replaced with the copper nanoparticles which serve as an affordable alternative with all the goodness and thus help in the production of highly efficient low-cost PSCs. These low-cost solar cells would serve as a major source of energy within the remote parts of the world. This study is further extended to the optimization of its different parameters, including radius, shape, and location of the particle embedded within the perovskite film of the solar cell. We aim to demonstrate the impact of different shapes and orientations of plasmonic nanoparticles embedded within the 200 nm perovskite film. The study is extended to nanospheres, nanocubes, nanocylinders, nanorods, and nanotriangular plates of varying sizes placed at the centre of the film. This geometric study also includes different orientations of a particle within the film, with respect to the source of light. To avoid plasmonic coupling occurring due to the close vicinity of two nanoparticles, its dimer formation is also taken account. We further extend our study to observe the anomalous effects on solar absorbance with the insertion of dielectric coated plasmonic metal nanoparticles within hybrid organic–inorganic halide perovskite solar cells. Insertion of the metal@dielectric core–shell nanoparticles in the active layers of thin-film organic solar cells enhances their conversion efficiency. The metal core increases the optical absorption, subsequently the conversion efficiency of thin-film organic solar cells due to LSPR-based field enhancement effect; the dielectric shell plays a key role to keep the metal core inaccessible so that it cannot offer recombination centre of the light-induced excitons.

2 Designing of PSC

The systematic model of designed solar cell embedded with plasmonic nanoparticle is shown in Fig. 1. Here, Spiro-OMeTAD (ns) chip acts as a hole transport material (HTM), while semi-infinite glass layer (n_g) of thickness 100 nm acts as a substrate. Moreover, the refractive index and the extinction coefficient of the materials utilized are extracted from Sahai et al. (2021); Sahai and Varshney 1088; Johnson and Christy 1972). To reduce the toxicity due to high concentration of lead, the thickness of the perovskite layer (n_p) containing the plasmonic nanoparticle is optimized to 200 nm. The geometrical parameters of the perovskite layer and nanoparticle are considered as ($L_X \times L_Y \times L_Z$) and radius ($r$), respectively.

To evaluate the performance of the plasmonic PSCs, we have employed finite difference time domain method (FDTD).A plane source of light, travelling in the z-direction, is irradiated on the Spiro-OMeTAD chip perpendicularly. Assuming a semi-infinite structure, periodic boundary conditions are applied in the x-direction and y-direction. Perfectly matched layer (PML) is applied in the z-direction to minimize optical losses. High mesh density of 5 nm is applied across the perovskite film containing nanoparticle, to achieve high convergence of results.

Finally, to quantify the performance of the PSCs, with and without embedded nanoparticle, we determine the total solar absorption given by Zhang et al. (2015); Omelyanovich et al. 2016; Ghahremanirad et al. 2017).

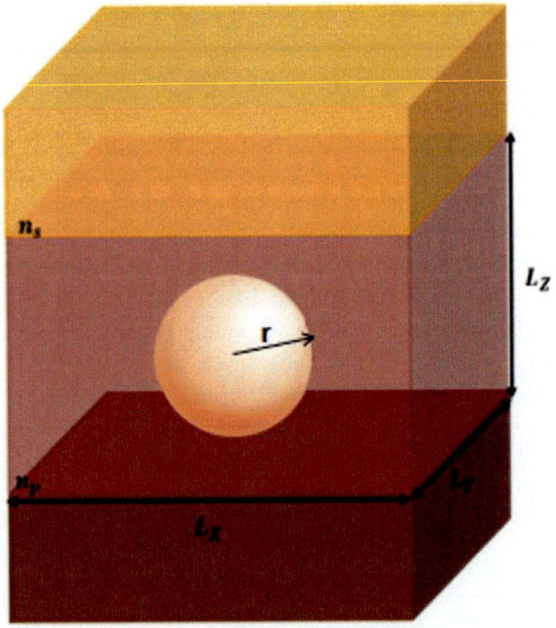

Fig. 1 Perovskite solar cell ($L_X \times L_Y \times L_Z$) containing nanoparticle of radius r

$$P_{abs} = \int \varepsilon_0 \omega(\lambda) |E(x, y, z, \lambda)|^2 n(\lambda) k(\lambda) \mathrm{d}V$$

where ω is the angular frequency, *E* is the electric field, $n(\lambda)$ and $k(\lambda)$ are the spectral dependent real and imaginary part of the material over whose volume *V*, and the integral is numerically calculated.

3 Material of the Nanoparticles

In this section, we examine the consequence of different metallic nanoparticles, e.g. gold, silver, and copper, embedded within the perovskite layer of 200 nm thick. The solar absorbance of perovskite film in the absence of a nanoparticle is kept as a reference. These metallic spherical nanoparticles of constant volume of radius 60 nm are placed at the middle of the film. It is quite apparent from Fig. 2 that copper nanoparticles exhibit parallel enhancement as silver and gold nanoparticles. Strong local fields are produced by the particle which enhances the overall absorption efficiency of the cell (Perrakis et al. 2019).

The optical properties of these particles depend on their profile and dielectric medium in which it is placed. Mie theory gives the analytical solution of Maxwell's equations as the incident light is scattered by the particle of any size (Qiang et al. 2001; Wriedt 2012; Bohren and Huffman 1998). This theory is confined to spherical nanoparticles but can be modified for other shapes (Bohren and Huffman 1998).

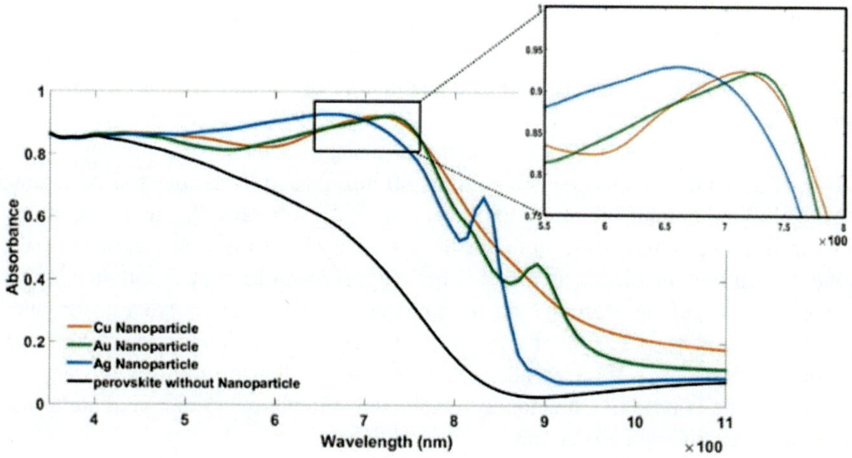

Fig. 2 Solar absorbance enhancement with plasmonic nanoparticles. The inset indicates that copper nanoparticle (orange) portrays similar enhancement as gold (green) and silver (blue) (Sahai et al. 2021)

According to this theory, absorption cross section $Q_{\text{absorption}}$ can be defined as the difference between the extinction cross section $Q_{\text{extinction}}$ and scattering cross section $Q_{\text{scattering}}$ given by

$$Q_{\text{absorption}} = Q_{\text{scattering}} - Q_{\text{extinction}}$$

The calculated extinction cross section and scattering cross section are given by,

$$Q_{\text{extinction}} = \frac{2\pi}{k^2}\sum_{p=1}^{\infty}(2p+1)\{a_{\text{p}} + b_{\text{p}}\}$$

$$Q_{\text{scattering}} = \frac{2\pi}{k^2}\sum_{p=1}^{\infty}(2p+1)\left(|a_p|^2 + |b_p|^2\right)$$

In these set of equations, a_{p} and b_p are the scattering coefficients. k is defined as the wave number given by $k = 2\mu p/\lambda$ where μ is the refractive index of the medium in which the particle is entrenched.

Where the scattering coefficients a_p and b_p are given as,

$$a_p = \frac{q\Psi_p(qx)\Psi'_p(x) - \Psi_p(x)\Psi'_p(qx)}{q\psi_p(qx)\xi'_p(x) - \xi_p(x)\psi'_p(qx)}$$

$$b_p = \frac{\Psi_p(qx)\Psi'_p(x) - q\Psi_p(x)\Psi'_p(qx)}{\psi_p(qx)\xi'_p(x) - q\xi_p(x)\psi'_p(qx)}$$

In these equations, q is the relative refractive index, x is defined as the size parameter and is given $x = k\,d/2$ where d is the diameter of the particle. Ψ_p and ξ_p are the Bessel functions which can be deduced as the ratio of extinction factor to the cross section of the particle.

Figure 3a illustrates Q_{ext} for diverse metallic nanoparticles like copper, gold, and silver of definite size with constant radius 60 nm, placed in the perovskite region. These plots are obtained from Philip Laven (MiePlot) software. Q_{ext} indicates strong localized and scattered field around the particle. The Q_{ext} is the vital parameter to the solar absorption enhancement which is quite apparent from Figs. 2 and 3a. Copper and gold nanoparticles demonstrate an equivalent red shift in the solar absorbance wavelength and Q_{ext}. The solar absorbance is maximum at longer wavelengths ($\lambda > 600$ nm) which covers a broader spectral range of the infrared region. Results reveal that the copper is a financially viable and suitable material candidate to use for enhancing the light entrapping ability of PSCs.

The enclosed dielectric medium is also an important facet to study the optical properties of a plasmonic nanoparticle. Oxidation is the major problem in the proposed use of the aforementioned particle. To avoid the oxidation of copper in the air, we must place it in the dielectric medium which has higher refractive index than air.

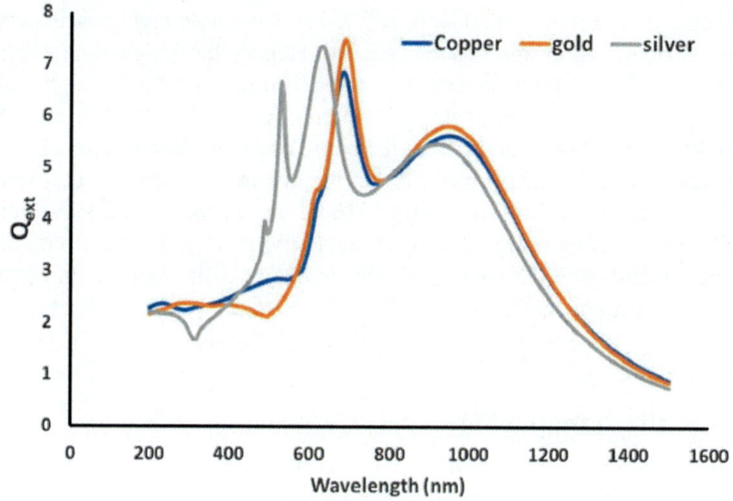

(a)

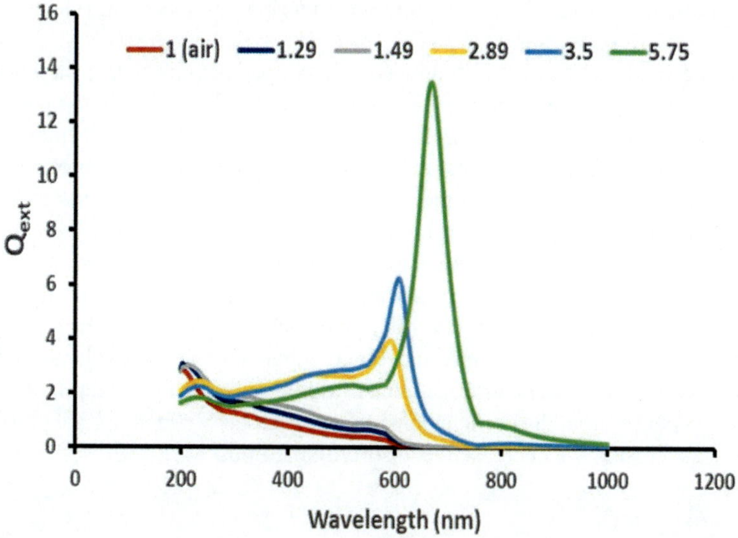

(b)

Fig. 3 **a** Q_{ext} spectrum for Cu, Ag, and Au nanoparticles of radius 60 nm radius implanted within perovskite. **b** The extinction cross section for copper nanoparticle implanted in different dielectric mediums. (Sahai et al. 2021)

The dielectric constant of perovskite (ε = 5.75) is reasonably high; hence, the probability of its oxidation is significantly reduced. Hence, the designed model can be implemented if the copper nanoparticle is embedded within a film of higher dielectric medium. Figure 3b authenticates this finding; it describes Q_{ext} for a Cu-NP of diameter 40 nm, positioned in different dielectric mediums ranging from air (ε = 1) to perovskite (ε = 5.75). It is observed that the due to its oxidation, Q_{ext} weakens for the dielectric constant values close to 1. Thus, we can approve the substitution of gold and a silver nanoparticle with copper nanoparticles. Cu-NP significantly lowers the production costs of PSCs and makes it an economical alternative to be employed for commercial purposes.

4 Size of the Nanoparticle

The size of the nanoparticle is very important to optimize the solar absorbance at longer wavelengths. We may study this feature by considering the range of the radius of the particle placed in the perovskite layer. The solar absorbance increases as the radius (*R*) of the particle increases, due to scattering enhancement (George et al. 2019).

With the incorporation of different size of copper nanoparticles, the enhancement in solar absorbance is quite visible in Fig. 4a.The absorption of the perovskite layer without any particle is kept as a reference to measure the enhancement factor. Maximum solar absorption is obtained due to the double contribution of near- and far-field effects with the incorporation of nanoparticle of radius 70 nm. This NP is placed at the centre of the 200-nm-thick perovskite film. The enhanced electric (***E***) field has been observed in Fig. 4b due to the presence of 70 nm plasmonic nanoparticle. With the presence of these plasmonic particles, an enhancement of the light trapping ability of the PSCs is observed at longer wavelengths (λ > 600 nm). Particles having radius > 70 nm do not offer adequate enhancement; the explanation is the ohmic losses and the large near-field effects. Additionally, the light trapping ability of the PSCs depends upon the particle density; it reduces as we increase the plasmonic particles; as a result, we decrease the absorbing medium.

5 Location of the Nanoparticle

The deepness of the nanoparticle in the perovskite thin layer is a further crucial aspect in the manipulating of the solar absorbance of PSCs. The position of the nanoparticles is varied along the z-direction. The radius (70 nm) of the particle and the dimension of the cell ($L = 200$ nm) are kept constant. The study has been conducted at three distinct vertical positions, (I) the particle is placed at the centre, case (II) the particle is placed at $z = +20$ nm (shifted towards the source), and case (III) the particle is placed at $z =$ -20 nm (20 nm towards the substrate). From Fig. 5, it can be accomplished

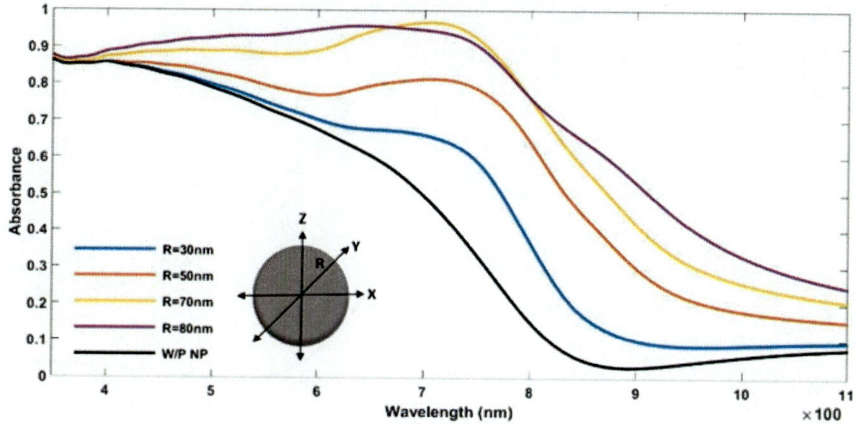

(a)

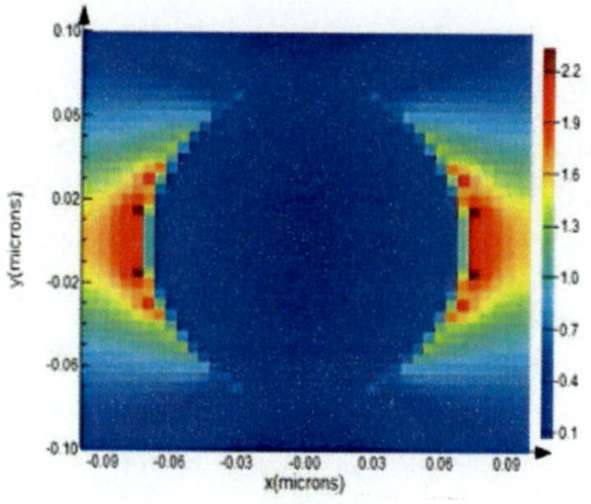

(b)

Fig. 4 **a** Variation of solar absorbance with different size copper nanoparticles rooted in perovskite layer; **b** electric field of Cu-NP ($R = 70$ nm) placed in the perovskite layer (Sahai et al. 2021)

that we get the most satisfactory result when the particle is placed at the middle of the perovskite film; it is due to the double contribution of both near and scattering effects of the plasmonic nanoparticle. For case (II), scattering effects are concealed by the far-field effects, while for case (iii), scattering effects preponderate in the cell which leads to the reduction of solar absorbance.

6 Shape of the Nanoparticle

The LSPR of metallic nanoparticles also depends on their morphology. Here, we critically analyse the different shapes of copper nanoparticles, e.g. sphere, ellipsoid,

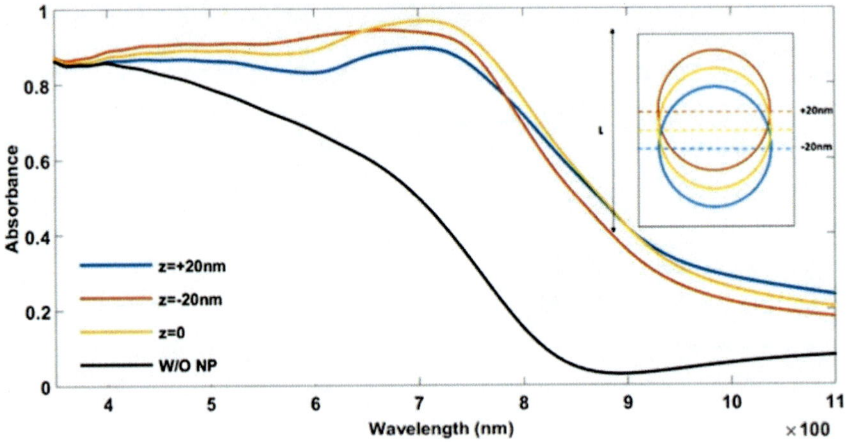

Fig. 5 Solar absorbance with varying location of nanoparticle in perovskite film. (Sahai et al. 2021)

cube, bars, rod, etc. entrenched in the thin perovskite layer of PSCs. This study is further extended to different orientations of these plasmonic nanoparticles with respect to the incident light, which influence the solar absorbance enhancement of the cell.

6.1 *Nanoshpere*

A spherical copper nanoparticle of varying radius is embedded within the perovskite layer of thickness 200 nm. The particle is placed at the centre of the layer to obtain maximum solar absorption enhancement. The localized surface plasmon resonance (LSPR) of a particle depends vitally upon the size of the particle. Mie theory (Ghahremanirad et al. 2017; Qiang et al. 2001; Wriedt 2012) can be utilized to explain this optical property of a spherical particle. With the increase in the size of the particle, the extinction efficiency of the particle increases. Figure 6 depicts the absorbance spectrum with the incorporation of different sizes of spherical nanoparticles. The black curve denotes the absorbance of perovskite layer without the presence of a plasmonic nanoparticle. It is kept as a reference to testify the influence of varying sizes of copper nanoparticles. We observe that with the increase in the radius of the particle from 30 to 70 nm, the light trapping ability of PSCs increases simultaneously due to the near- and far-field scattering effect. The increase in the radius of the particle leads to the scattering enhancement within the perovskite matrix. Moreover, we also observe enhanced solar absorbance at longer wavelengths ($\lambda > 650$ nm). We also notice that the size of the particle exceeded beyond 70 nm, i.e. 80 nm, and the absorbance efficiency of the PSC decreases sharply. Though, larger particles are associated with the small ohmic losses and large near-field effect, they do not necessarily

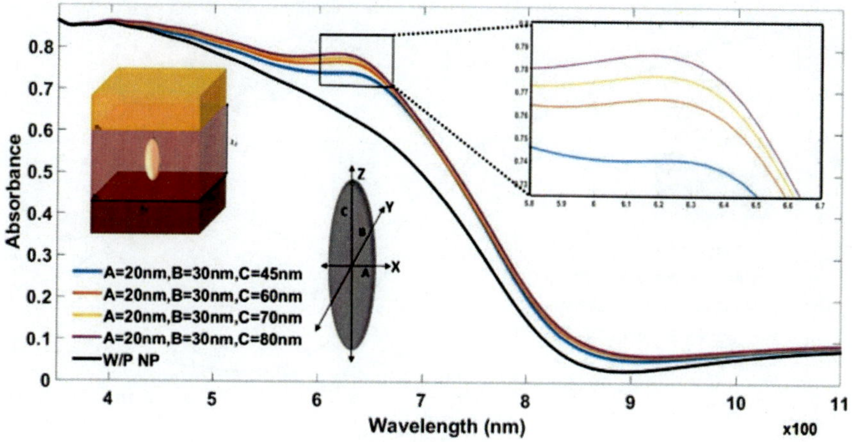

Fig. 6 Solar absorbance enhancement with the presence of prolate nanoparticle (Sahai et al. 2021)

offer enhanced absorbance in PSC, as they replace large volumes of the absorbing material which ultimately impacts the overall light trapping ability of the cell. Thus, we can conclude that spherical particles of size 70 nm incorporated within the cell lead to maximum solar absorbance along with minimum ohmic losses. The enhanced electric field (E) of the perovskite film containing the 70 nm copper nanoparticle can be attributed to the LSPR, characteristic property of metallic nanoparticles.

6.2 Nanoellipsoid

Localized surface plasmon resonance (LSPR) of nanoparticles depends not only upon its dimension and shap but also upon its inclination in the perovskite thin film with respect to the direction of the incident light. Nanoellipsoids demonstrate an ideal instance among metallic nanoparticles describing transverse and longitudinal resonance. Here, we would critically analyse two models connected to it, naming oblate and prolate particles.

6.3 Prolate-Ellipsoid

Prolate indicates to the ellipsoidal nanoparticle when it is placed parallel to the direction of propagating incident light. The prolate nanoparticle is placed in the middle ($z = 0$) of the source and the substrate. The geometrical parameters of the particle refer to A nm (along x-axis), B nm (along y-axis), and C nm (along z-axis). The size of the particle can be tailored by modifying C parameter, which elongates

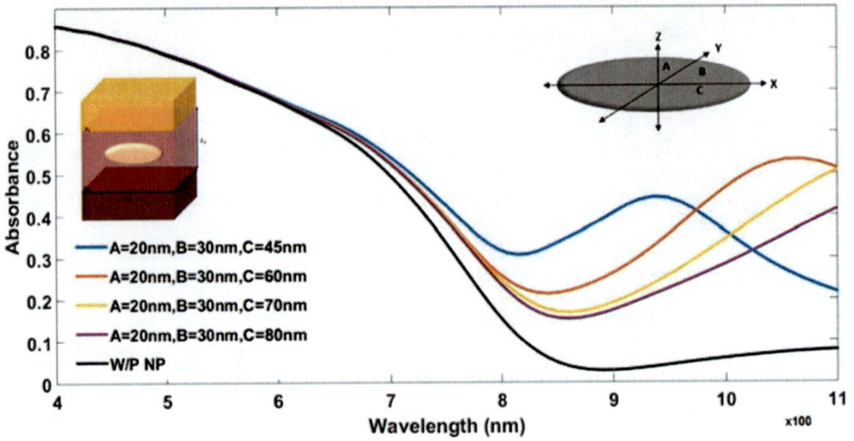

Fig. 7 Solar absorbance with the incorporation of oblate copper nanoparticle within the perovskite film (Sahai et al. 2021)

the particle along *z*-axis (along the incident light). AS a consequence, it leads to the variation in the solar absorbance enhancement. Figure 6 shows the solar absorption curve which concludes that, though the elongation of the particle does not lead to a significant increase in the enhancement factor of the solar absorbance, but the average of the spectra associated with its orientation within the perovskite layer is similar to the solar absorbance of spheres.

6.4 *Oblate-Ellipsoid*

For oblate, ellipsoidal nanoparticle is placed perpendicular to the direction of light. As the orientation of a plasmonic nanoparticle associated with the propagating incident light influences its LSPR wavelength, a red shift in solar absorbance wavelength ($\lambda > 800$ nm) is observed with the increase in the transverse length of the particle (C), along the *z*-direction. Figure 7 reveals that the enhancement factor of solar absorbance is significantly reduced with the increase in the size of an oblate nanoparticle. It can be associated with the minimized the near-field effect produced by the particle.

6.5 *Nanocylinder*

LSPR of a particle also depends upon its geometrical parameters. Nanocylinders set a perfect example for its illustration. Here, we have varied the height and the radius of the nanocylinders, placed parallel to the direction of incident light. The particle is placed at $z = 0$, i.e. at the centre of the film. The black curve in Figs. 8

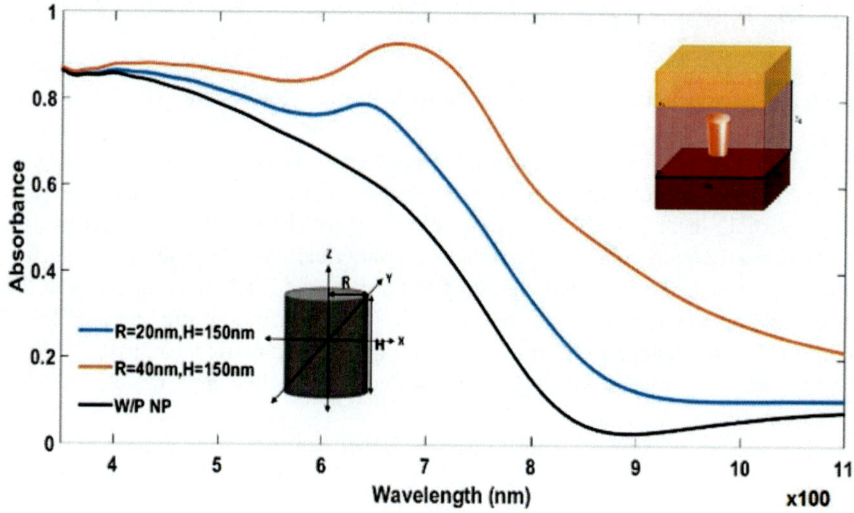

Fig. 8 Solar absorbance of perovskite layer with different radius of nanocylinders keeping its height constant at 150 nm (Sahai and Varshney 1088)

and 9 depicting perovskite absorbance of the PSCs without any particle is kept as a reference for observing the enhancement in solar absorbance with inclusion of a particle.

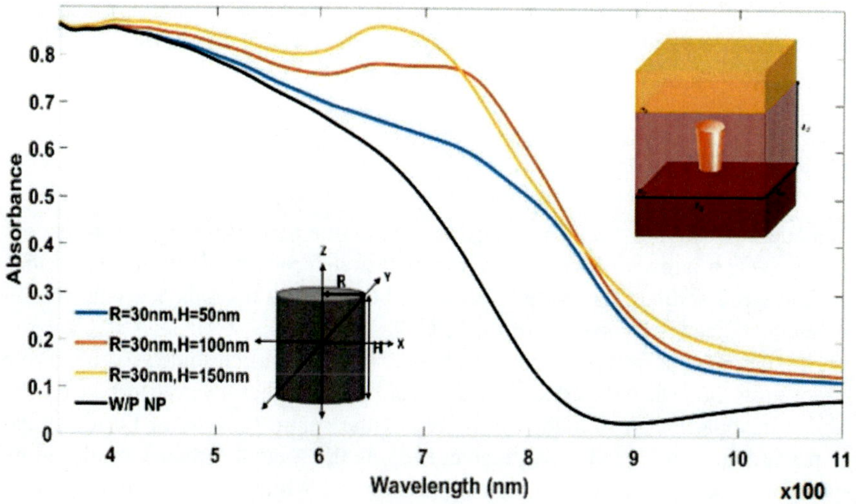

Fig. 9 Solar absorbance of perovskite layer with different heights of nanocylinders, keeping its radius constant at 30 nm (Sahai and Varshney 1088)

6.6 *Radius of the Cylinder*

The performance of the PSCs is evaluated with the inclusion of nanocylinders, whose radius is varied from 20 to 40 nm. The height of the cylinder is kept constant at 150 nm. In an ideal case, a nanocylinder behaves as nanorod, displaying LSPR in transverse and longitudinal directions. From Fig. 8, we observe that as the radius of the particle increases, the absorption efficiency of the cell increases but overall absorption efficiency remains close that of a spherical particle. The scattering far-field effect induced by the particle along the transverse direction contributes to the absorbance enhancement. It can be concluded that absorption efficiency of PSCs embedded with a nanoparticle largely depends upon its geometrical parameters. Note, the absorbance wavelength is slightly red shifted towards the longer wavelengths. ($\lambda > 650$ nm).

6.7 *Height of the Cylinder*

LSPR, the characteristic property of metallic nanoparticles, depends upon various parameters. On increasing the height of the nanocylinder from 50 to 150 nm and keeping its radius constant at 30 nm, it behaves as an elongated particle placed parallel to the direction of the incident light as shown in Fig. 9. Interestingly, though the absorption efficiency of the cell enhances with the increase in the height of the nanocylinder, the absorbance wavelength blue shifts towards the visible wavelength. Hence, an optimum balance must be maintained between the height and radius of nanocylinder to enhance the solar absorbance of the PSCs at longer wavelengths.

6.8 *Nanocubes*

Nanocubes of varying sizes are embedded within the perovskite layer of thickness 200 nm. Due to greater charge separation, the resonance wavelength for nanocubes is slightly red shifted. The corner sharpness of the edges also plays a vital role in enhancing the solar absorbance of the PSCs.

The sharpness of the corner edges in plasmonic nanoparticles plays a crucial role in elevating the near-field effect and thus leading to higher solar absorbance. The cubical nanoparticle with side A is placed at the centre ($z = 0$) of the film, and the performance of PSCs is investigated, keeping the solar absorbance produced by perovskite layer without nanoparticle as a reference. With the increase in the size of the nanocube, a red shift in solar absorbance wavelength ($\lambda > 700$ nm) is observed as compared to spherical nanoparticles, which portray absorbance at wavelengths close to the visible region. Though, the average enhancement factor for both spherical and cubical nanoparticle remains the same. Figure 10b and c depict enhanced E field

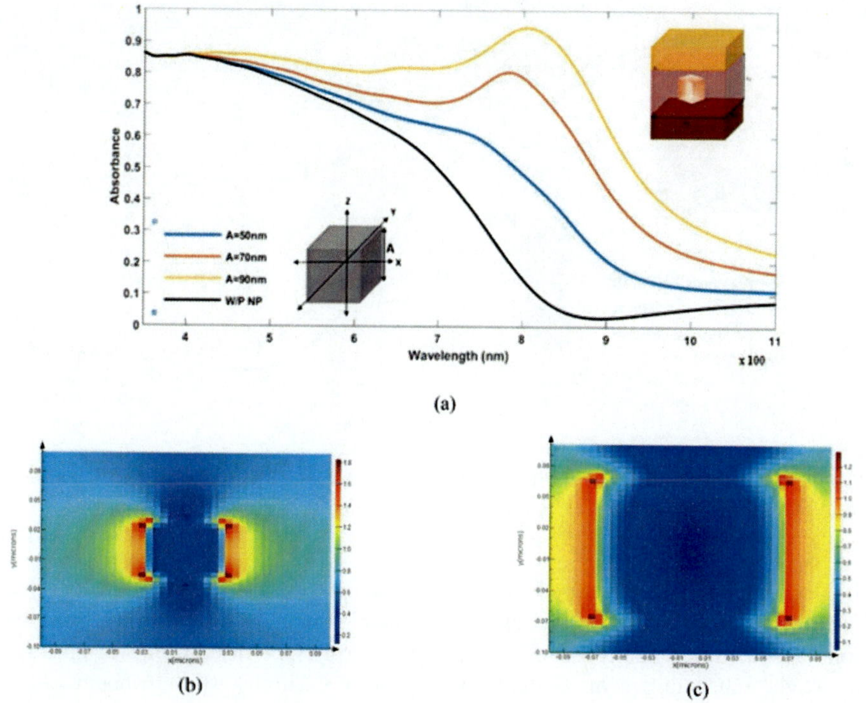

Fig. 10 **a** Solar absorbance enhancement due to incorporation of different sizes of nanocubes. Electric field enhancement due to **b** 50 nm **c** 70 nm (Sahai et al. 2021)

with the inclusion of 50 and 70 nm particle, respectively, thus supporting intensified solar absorbance observed in Fig. 10a.

6.9 Nanobars

Nanobars display LSPR along transverse and longitudinal directions. Here, we have placed the nanoparticle at the centre of the perovskite film of thickness 200 nm and investigated the performance of the PSC depending upon its orientation with respect to the incident light. The corner sharpness induced by the edges of the nanobar also plays a vital role in enhancing the solar absorbance.

6.10 Parallel to the Direction of Incident Light

Copper nanobar, placed parallel to the direction of incident light, is embedded within the perovskite film. The height of the nanobar is varied along the z-direction from

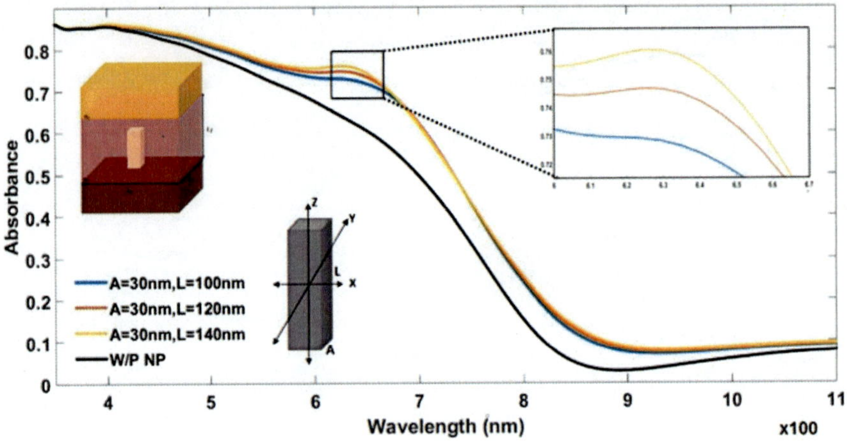

Fig. 11 Solar absorbance of perovskite layer with nanobar placed parallel to the direction of incident light (Sahai and Varshney 1088)

100 to 140 nm. The length and breadth of the nanobar are assumed to be *A*. Its value fixed at 30 nm. We observe that as the particle elongates along the *z*-direction, the solar absorbance of the cell does not increase sharply. The inset in Fig. 11 portrays the slight enhancement in its light trapping ability. This can be attributed to the LSPR along the direction of incident light. Note, no shift in absorbance wavelength is observed along with elongation of the particle.

6.11 *Perpendicular to the Incident Light*

Copper nanobar placed vertically within the perovskite film lies perpendicular to the direction of incident light. Nanobar with similar geometrical parameters is placed within the perovskite film, and its solar absorbance is investigated. Here, we observe that the overall solar absorbance enhancement of the designed solar cell does not increase sharply with the elongation of the particle along the *x*-direction as shown in Fig. 12. The absorbance wavelength lies within the visible range of the electromagnetic spectrum and no shift is observed with the change in its geometrical parameters. Note, the overall enhancement in the light trapping ability of the perovskite film embedded with a copper nanobar, placed parallel to the direction of light, portrays a significant enhancement as compared to the particle placed perpendicular to the direction of the incident light.

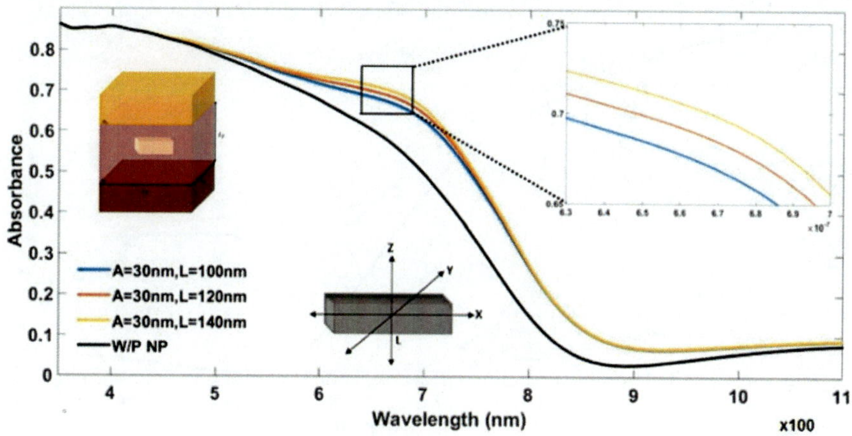

Fig. 12 Solar absorbance of perovskite layer with nanobar placed perpendicular to the direction of incident light (Sahai and Varshney 1088)

6.12 *Triangular Nanoplate*

Equilateral triangular nanoplate of side *A* and thickness *T* is embedded within the perovskite film of the designed solar cell.

The thickness of the plate is kept constant at 30 nm. The side of these equilateral triangular plates is varied from 100 to 140 nm to investigate the performance of the designed solar cells. The black curve in Fig. 13 represents the solar absorbance of the perovskite film without the presence of a nanoparticle. Interestingly, we observe that

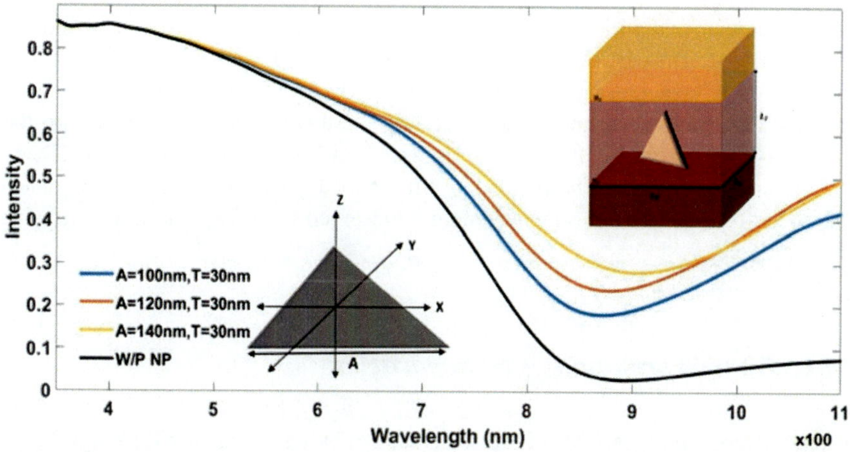

Fig. 13 Solar absorbance of perovskite layer with triangular nanoplate placed within the perovskite film (Sahai and Varshney 1088)

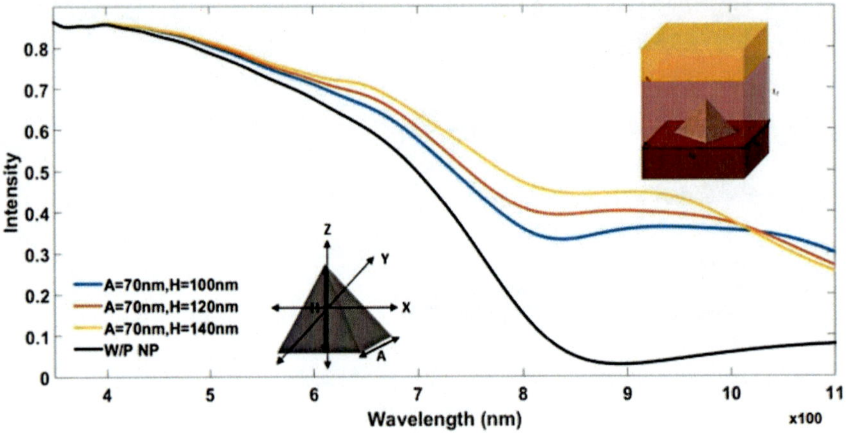

Fig. 14 Solar absorbance of perovskite layer with rectangular pyramid placed within the perovskite film (Sahai and Varshney 1088)

with the inclusion of these particles at the centre of the film, only a slight enhancement in its solar absorbance is achieved. The geometrical parameters and morphologies of these particles play a quite significant role in determining its absorbance efficiency. Note, no shift in its absorbance wavelength is observed with the increase in the size of the particles.

6.13 Rectangular Pyramids

The rectangular pyramid with square base A and height H is embedded within the perovskite film. The particle is placed at the centre ($z = 0$) of the film. The height of the pyramid is varied along z-axis to investigate the light trapping ability of PSCs. Minimal enhancement in the solar absorbance of these cells is observed with the elevation in the height of the particle as shown in Fig. 14. The overall efficiency remains similar to that of the perovskite film embedded with triangular nanoplates and no shift in the absorption wavelength is observed with the variation in the size of the particle.

6.14 Dimer Formation of Nanoparticles

Plasmonic coupling (Carretero-Palacios et al. 2015) is associated with a significant red shift in the resonance wavelength as electron cloud of a particle hinders the electron cloud of another when placed in the proximity. To minimize this effect, a constant volume of the unit cell is considered for calculating the optimized distance

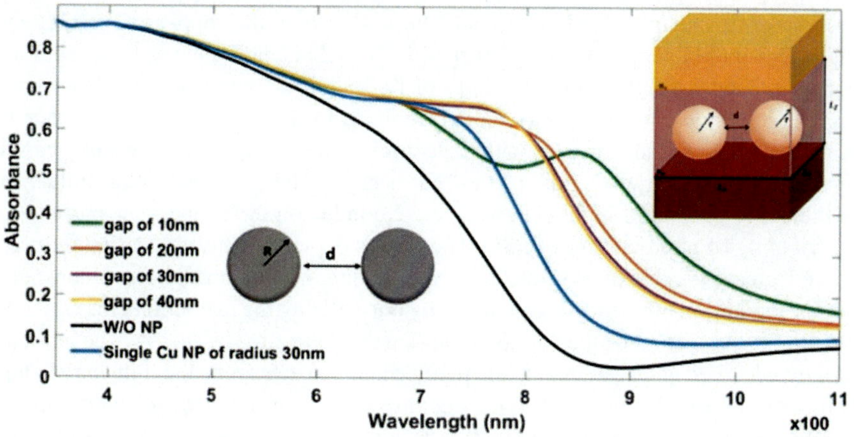

Fig. 15 Solar absorbance with the incorporation dimer of nanoparticles ($R = 30$ nm) within the perovskite film (Sahai et al. 2021)

(d) between two plasmonic nanoparticles of radius 30 nm. The study has been carried out by altering the separation between the plasmonic particles.

The plot shown in Fig. 15 highlights a large red shift in the solar absorbance with the increase in the distance between two spherical copper nanoparticles beyond 30 nm. Hence, we mark 30 nm as the safe distance for the placing of these particles within PSCs, to avoid a red shift occurring due to plasmonic coupling. Interestingly, it also noted that two nanoparticles of same size produce higher solar absorbance as compared to a single nanoparticle placed at an optimized distance to avoid plasmonic coupling. Hence, usage of larger number of smaller nanoparticles is encouraged to enhance the solar absorption of PSCs.

6.15 *Core Shell Plasmonic Nanoparticle*

In this section, we study the anomalous effect of dielectric coated plasmonic Cu-nanoparticle on solar absorbance enhancement when placed in the hybrid organic–inorganic halide perovskite solar cell. The insertion of the metal@dielectric core–shell nanoparticles in the active layers of thin-film organic solar cells affects their conversion efficiency. The metal core increases the optical absorption and consequently the conversion efficiency of thin-film organic solar cells due to LSPR-based field enhancement effect and meanwhile the dielectric shell avoids the metal core to offer recombination centre of the light-induced excitons.

Studies carried out in the past to enhance the performance of PSCs with the introduction of core–shell are done using expensive metallic nanoparticles, e.g. silver or gold. These nanoparticles are definitely stable and provide very good results but we must explore some low-cost yet effective substitutes of nanoparticles. In the previous sections of this chapter, we have already discussed about the copper as an excellent replacement of silver and gold (Sahai et al. 2021; Sahai and Varshney 1088) without compromising the light trapping ability of PSCs. On the basis of the inference of our previous study, we present a detailed theoretical study to elucidate the impact of core–shell copper nanoparticles with dielectric coatings. A comparative study based on finite difference time domain method has been conducted to analyse the influence of dielectric coating of high and low refractive index on the copper nanoparticle in the yielding of photocurrent of PSC. The results depict that the absorption reduces as we increase the thickness of the dielectric coating, keeping the fill factor constant.

The extinction efficiency is calculated by normalizing the extinction cross section over the cross section of the nanoparticle (Du 2004) using Mie theory. The amount of light adsorbed and scattered by the particle helps to determine the light harvesting ability of PSCs embedded with nanostructures over a particular wavelength range.

The designed perovskite solar cell is irradiated with a plane wave as shown in Fig. 16, propagating along the z-direction whose intensity is normalized to match the standard solar power spectrum. Periodic boundary conditions are applied in the x-direction and y-direction while a perfectly matched layer (PML) is applied in the z-direction to prevent further optical losses. A high mesh density of 5 nm is applied to achieve elevated accuracy and convergence of results.

The simulation results show the absorbance spectra and electric field intensity produced by particles with dielectric coating of lower refractive index increases

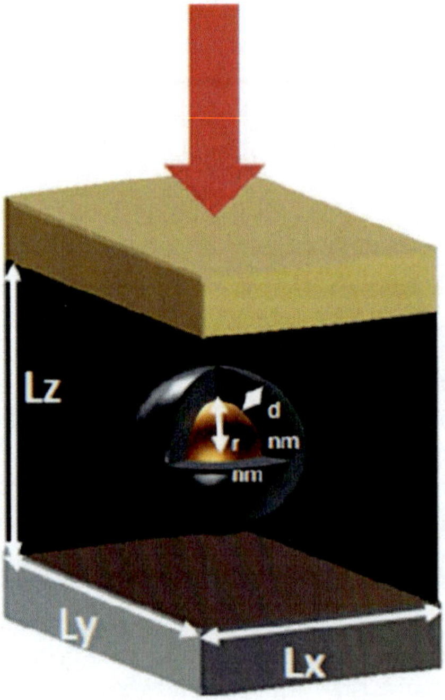

Fig. 16 Schematic of unit cell PSCs embedded with a plasmonic copper nanoparticle with the core radius r nm and the depth of the shell *d* nm

significantly as the shell width shrinks. These aberrant effects expand our knowledge on plasmonic PSCs and thus open the doors for its optimized design with enhanced light trapping ability. On comparison with different available core metals like Au, Ag, and Cu, it is being observed that the copper@dielectric offers chemical and thermal stability with increased absorption. It further lowers down the production costs these designs, enabling them to be utilized even within the underdeveloped countries.

When the dielectric shell is introduced, its refractive index plays a key role in determining the enhancement in its absorbance efficiency. Figure 17a portrays the extinction spectrum of nanostructures placed in perovskite medium, with different thickness of core and shell thus keeping the total size of the particle constant at 70 nm. It is evident from the graph the increase in the shell thickness from $d = 0$ nm to $d = 15$ nm, and a simultaneous decrease in the size of the core from $r = 70$ nm to $r = 55$ nm, respectively, the extinction efficiency decreases. The results portrayed by Fig. 17b further justify the above argument, the absorbance intensity of Cu@SiO_2 decreases with the enhancement of the shell thickness.

Table 1 shows the J_{sc} of these core–shell nanostructures. It is utilized to examine the absorption efficiency of these cells. The J_{sc} for PSCs without any nanoparticle is found to 17.8421 mA/cm^2. The enhancement factor is calculated using

$$\text{Enhancement} = \frac{J_{\text{SC, with NP}}}{J_{\text{SC, with out NP}}}$$

Fig. 18 reveals the reduction in the enhancement factor is observed for Cu@SiO_2 embedded within the perovskite film.

The electric field intensity is governed by the boundary conditions at the interface of the particle. Figure 19a represents the electric field intensity of Cu@SiO_2 with a core of 69 nm and a shell of 1 nm. As the thickness of the shell increases to 15 nm, the electric field intensity as shown in Fig. 19b decreases significantly. Note, the electric field decreases exponentially as we move away from the particle interface.

The dielectric shell provides chemical stability to the particle embedded within the perovskite film by preventing the oxidation of the copper nanoparticles and thus help maintain the efficiency of these devices for a longer duration of time.

7 Summary

In conclusion, this work mainly focused on the study of solar absorbance enhancement dependency on the geometry and material of plasmonic nanoparticles rooted within organic–inorganic halide perovskite solar cells (PSCs). The choice of the nanoparticle's material is relative to the enhancement factor of the cell. Copper nanoparticle proves its candidature in PSC designs, with some intriguing advantage over gold and silver. It is observed that the copper significantly lowers the effective cost without negotiating the solar absorbance of the cell. The volume and the position of the particle within the 200-nm-thick perovskite film are optimized to enrich

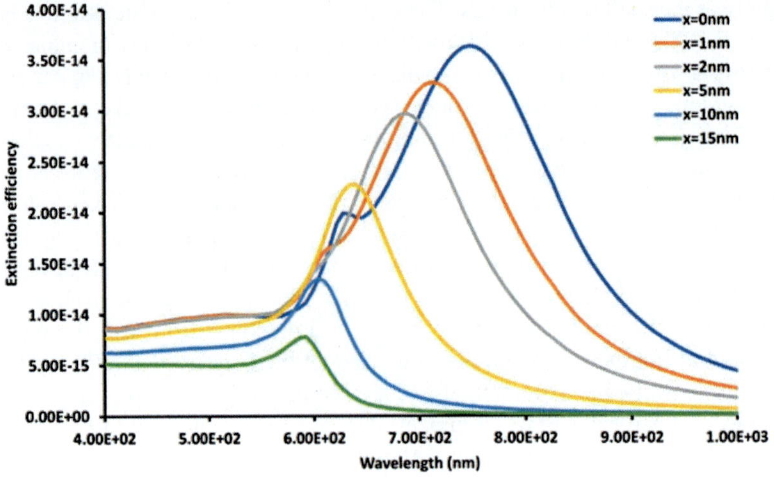

(a)

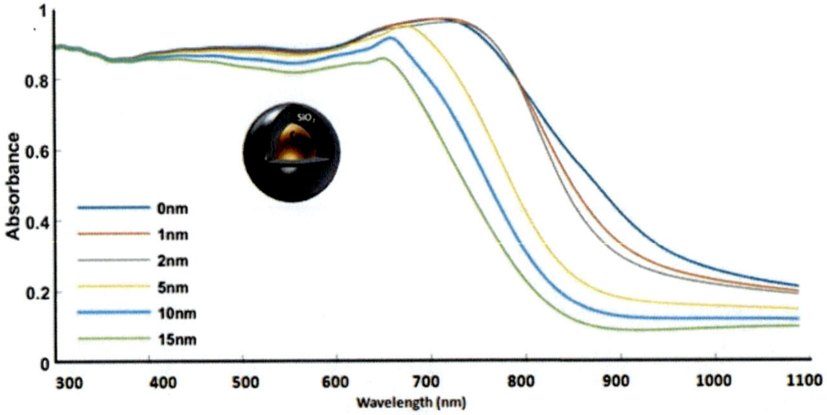

(b)

Fig. 17 **a** Extinction efficiency of Cu@SiO2. **b** Absorbance produced by Cu@SiO2 within the perovskite film. The thickness of the core and shell are varied from 70 to 55 nm and 0 to 15 nm, respectively

the solar absorbance of the designed solar cell. Results represent that the highest enhancement can be achieved with the insertion of spherical nanoparticles of 70 nm radii, positioned at the centre of the film. In this chapter, we highlight the impact of customized plasmonic nanoparticles including nanospheres, nanocubes, nanocylinders, nanorods, nanotriangular plates, etc. embedded within the film. This study based on the shape of nanoparticle is extended to the orientations of a particle within the film, with respect to the direction of the source of the light. It is observed that

Table 1 Calculation of the JSC and enhancement factor with the insertion of different Cu@SiO2 core–shell nanoparticles within thin perovskite film of PSCs

Size of the core (nm)	Size of the shell (nm)	J_{SC} mA/cm^2	Enhancement
70	0	31.0071	1.7378
69	1	30.219	1.6937
68	2	29.699	1.6645
65	5	26.088	1.4621
60	10	23.816	1.3348
55	15	21.737	1.2182

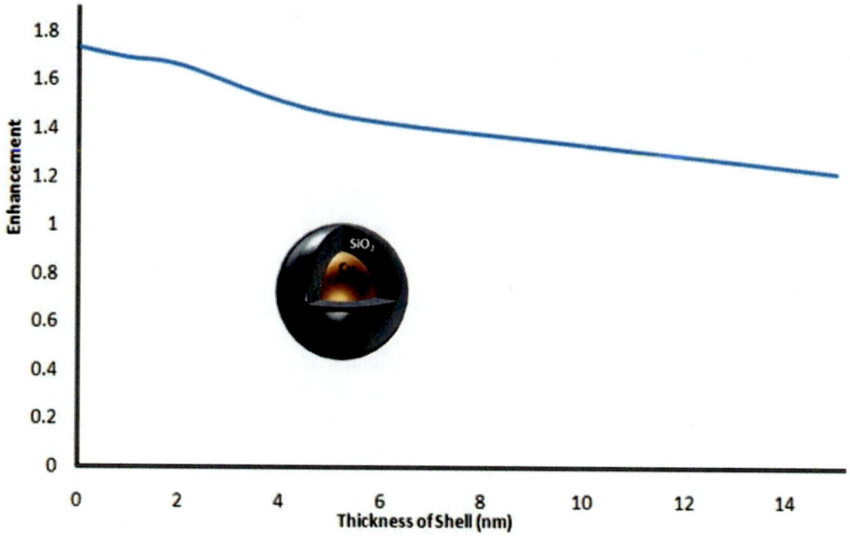

Fig. 18 Total enhancement in the light trapping ability of PSCs as a function of the shell thickness

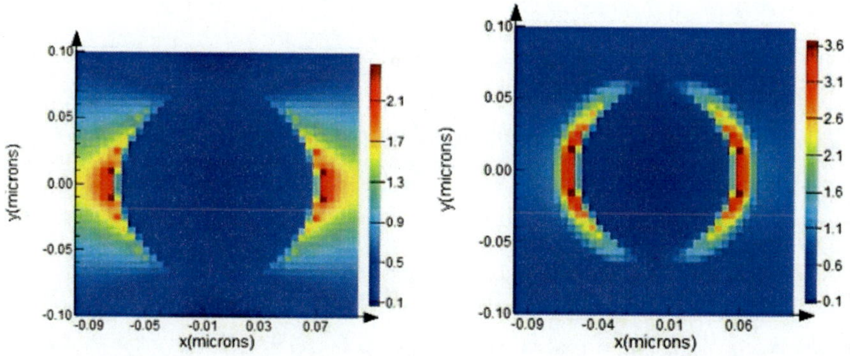

Fig. 19 Electric field intensity. **a** Cu@SiO2 with d = 1 nm, r = 69 nm. **b** Cu@SiO2 with d = 15 nm, r = 55 nm

the absorption efficiency of PSCs extremely depends upon the angle sharpness and point of reference of a nanoparticle within the film, though the average absorption remains comparable to the spherical particles.

To evade a red shift in the resonance wavelength due to plasmonic coupling, the dimer formation of nanoparticles is also taken into consideration. Optimization of safe plasmonic distance for two spherical nanoparticles embedded within the film is obtained to minimize this effect.

The anomalous effect on solar absorbance enhancement is also studied with the insertion of dielectric coated plasmonic copper nanoparticles within hybrid organic–inorganic halide perovskite solar cells. Conversion efficiency has been changed when we insert the metal@dielectric core–shell nanoparticles in the active layers of thin-film organic solar cells. Due to LSPR-based field amplification, metal cores boost optical absorption, improving the conversion efficiency of thin-film organic solar cells. The dielectric shell, on the other hand, keeps the light-induced exciton from coming into direct contact with the metal core, which would otherwise serve as a recombination centre.

References

M. Batmunkh, T.J. Macdonald, W.J. Peveler, A.S.R. Bati, C.J. Carmalt, I.P. Parkin, J.G. Shapter, Chemsuschem **10**(19), 3750–3753 (2017). https://doi.org/10.1002/cssc.201701056

C.F. Bohren, D.R. Huffman, Absorption and Scattering by a Sphere, in *Absorption and Scattering of Light by Small Particles* (John Wiley & Sons, Ltd, 1998a), pp. 82–129. https://doi.org/10.1002/9783527618156.ch4.

C.F. Bohren, D.R. Huffman, Absorption and Scattering by an Arbitrary Particle, in *Absorption and Scattering of Light by Small Particles* (John Wiley & Sons, Ltd, 1998b), pp. 57–81. doi: https://doi.org/10.1002/9783527618156.ch3

S. Carretero-Palacios, M.E. Calvo, H. Míguez, J. Phys. Chem. C Nanomater Interfaces **119**(32), 18635–18640 (2015). https://doi.org/10.1021/acs.jpcc.5b06473

S. Carretero-Palacios, A. Jiménez-Solano, H. Míguez, ACS Energy Lett. **1**(1), 323–331 (2016). https://doi.org/10.1021/acsenergylett.6b00138

J. Cui, C. Chen, J. Han, K. Cao, W. Zhang, Y. Shen, M. Wang, Adv. Sci. **3**(3), 1500312 (2016). https://doi.org/10.1002/advs.201500312

A. Dabirian, M.M. Byranvand, A. Naqavi, A.N. Kharat, N. Taghavinia, ACS Appl. Mater. Interfaces **8**(1), 247–255 (2016). https://doi.org/10.1021/acsami.5b08560

H. Du, Mie-scattering calculation. Appl. Opt. AO **43**(9), 1951–1956 (2004). https://doi.org/10.1364/AO.43.001951

R. Fan, L. Wang, Y. Chen, G. Zheng, L. Li, Z. Li, H. Zhou, J. Mater. Chem. A **5**(24), 12034–12042 (2017). https://doi.org/10.1039/C7TA02937C

Q. Fu, W. Sun, Appl. Opt. **40**, 1354–1361 (2001) https://doi.org/10.1364/AO.40.001354

E. Ghahremanirad, A. Bou, S. Olyaee, J. Bisquert, J. Phys. Chem. Lett. **8**, 1402–1406 (2017) https://doi.org/10.1021/acs.jpclett.7b00415

M.A. Green, A. Ho-Baillie, H.J. Snaith, Nat. Photonics **8**, 506–514 (2014). https://doi.org/10.1038/nphoton.2014.134

P.B. Johnson, R.W. Christy, Phys. Rev. B **6**(12), 4370–4379 (1972). https://doi.org/10.1103/PhysRevB.6.4370][CrossRef]

A.Z.Q. Luo, J. Shi, L. Yue, Z. Wang, X. Chen, S. Huang, Nanoscale **9**(8), 2852–2864 (2017). https://doi.org/10.1039/C6NR09972F

M. Omelyanovich, S. Makarov, V. Milichko, C. Simovski, Mater. Sci. Appl. **7**(12), 836–847 (2016). https://doi.org/10.4236/msa.2016.712064

N.K. Pathak, N. Chander, V.K. Komarala et al., Plasmonics **12**, 237–244 (2017). https://doi.org/10.1007/s11468-016-0255-9

G. Perrakis, G. Kakavelakis, G. Kenanakis, C. Petridis, E. Stratakis, M. Kafesaki, E. Kymakis, Opt. Express **27**, 31144–31163 (2019)

S. Sahai, A. Varshne, Opt. Spectro. **129**(8), 1088 (2021). https://doi.org/10.21883/OS.2021.08.51206.1008-21

S. Sahai, A. Varshney, Opt. Quant. Electron. **53**, 111 (2021). https://doi.org/10.1007/s11082-021-02755-9

M. Saliba, W. Zhang, V.M. Burlakov, S.D. Stranks, Y. Sun, J.M. Ball, M.B. Johnston, A. Goriely, U. Wiesner, H.J. Snaith, Adv. Funct. Mater. **25**(31), 5038–5046 (2015). https://doi.org/10.1002/adfm.201500669

T. Wriedt, M. Theory, A review, in *The Mie Theory. Springer Series in Optical Sciences,* ed. by W. Hergert, T. Wriedt, vol. 169 (Springer, Berlin, Heidelberg, 2012) https://doi.org/10.1007/978-3-642-28738-1_2

W. Zhang, M. Saliba, S.D. Stranks, Y. Sun, X. Shi, U. Wiesner, H.J. Snaith, Nano Lett. **13**(9), 4505–4510 (2013). https://doi.org/10.1021/nl4024287

W. Zhang, M. Anaya, G. Lozano, M.E. Calvo, M.B. Johnston, H. Míguez, H.J. Snaith, Nano Lett. **15**, 1698–1702 (2015). https://doi.org/10.1021/nl504349z][CrossRef]

Nanomaterials for Biomedical Engineering Applications

Anamika Singh and Dinesh K. Patel

Abstract Increasing complexity in the healthcare system has encouraged research toward novel prevention and treatment strategies. Over the period, nanotechnology emerged as a vital research area oriented toward different applications ranging from textile, defense, and chemical industry to biomedical fields. The process of knowing substances at the molecular level provoked the ascent of the innovative crucial and most promising research field of nanotechnology termed nanomedicine/nanobiotechnology. Researchers started investigating materials with exceptional properties that could be used in diagnostics, drug designing, drug delivery, and tissue engineering. The nanotechnology field has opened up the development of interestingly novel approaches in bioengineering by exploiting the unique properties of nanomaterials. These materials have many advantages in terms of surface area, physiochemical stability, non-toxicity, and biocompatibility compared to larger particles. The applications of nanomaterials have emerged in numerous technologies and devices such as diagnostic instrumentation, medical implants, and biocompatible devices for tissue regeneration or replacement, targeted delivery of drugs, hyperthermia, photoablation, cancer therapy, vaccines development, bioimaging, and biosensors.

Keywords Nanomaterials · Biomedical applications · Biosensors · Drug delivery · Nanotoxicology

---

A. Singh (✉)
Alexander Silberman Institute of Life Sciences, The Hebrew University of Jerusalem, Jerusalem, Israel
e-mail: anamika.iitr7@gmail.com

D. K. Patel
Human Computer Interaction Institute, Carnegie Mellon University, Pittsburgh, USA
e-mail: dineshpa@andrew.cmu.edu

J. K. Katiyar et al. (eds.), *Nanomaterials for Advanced Technologies*,
https://doi.org/10.1007/978-981-19-1384-6_5

1 Introduction

Nanomaterials (nanoparticles) are small particles (materials) having size < 100 nm. These materials have structural features between atoms/molecules and bulk materials. In the past, fabric dye, decorative glass, and treatment of arthritis were done using small metal particles. Romans and ancient Chinese used to embed the glass with different metal particles to generate dramatic color effects. Chinese porcelain also known as famille rose contain gold particles. Nanoparticles were studied systematically during the early seventeenth century (Huaizhi and Yuantao 2001). Along with small metal particles, silica particles also known as particles of insulators were used in number of applications which include automobile tires, optical fibers, catalyst support, etc. Similarly, magnetic oxides nanoparticles dispersed in different liquids (known as ferrofluids) have got interesting applications (Chikazumi et al. 1987). The famous lecture by the Nobel laureate Richard. P. Feynman in 1959 \where he stated, "*there is a plenty of room at the bottom,*" nanomaterial research interest was renewed (Feynman 2018) and discussed the potentiality of smaller materials. Off late large number of nanomaterials are synthesized, and their properties are being investigated and demonstrated. A new field called nanoscience and nanotechnology also emerged due to the very large number of studies being carried out on nanosized materials all over the world. A brief description is given below regarding the different types of nanomaterials that are commonly encountered.

1.1 Nanomaterial (NM) Classification

Classification of nanomaterials is based on the following criteria (Fig. 1) (i) dimensionality, (ii) morphology, (iii) state, and (iv) chemical composition (Pokropivny and Skorokhod 2007; Saleh 2020). Nanomaterials are classified as 0D, 1D, 2D, and 3D based on dimensionality. Nanomaterials having all their dimensions below 100 nm are termed as zero-dimensional (0D). Symmetric isotropic spheres, cubes, decahedrons, and tetrahedrons belong to 0D. Nanomaterials with one dimension above 100 nm and the other two below 100 nm are termed one-dimensional (1D). Nanowires, nanotubes, nanorods, nanofibers belong to 1D. Nanomaterials with two dimensions above 100 nm and one of the dimensions below 100 nm are termed as two-dimensional (2D). Disk and plates with polygon shapes belong to two-dimensional (2D). Nanomaterials with all dimensions beyond 100 nm are termed as three-dimensional (3D). Foams, fibers, nanopillars grew of substrates, electromechanical systems like NEMS and MEMS, polycrystals and 3D composites belong to three-dimensional (3D).

Based on morphology, they are classified as flatness, sphericity, and aspect ratio. Based on aspect ratio, it can be divided into two categories: low-aspect ratio and high-aspect ratio. Nanospheres, nanopyramids, and nanocubes belong to low-aspect ratio,

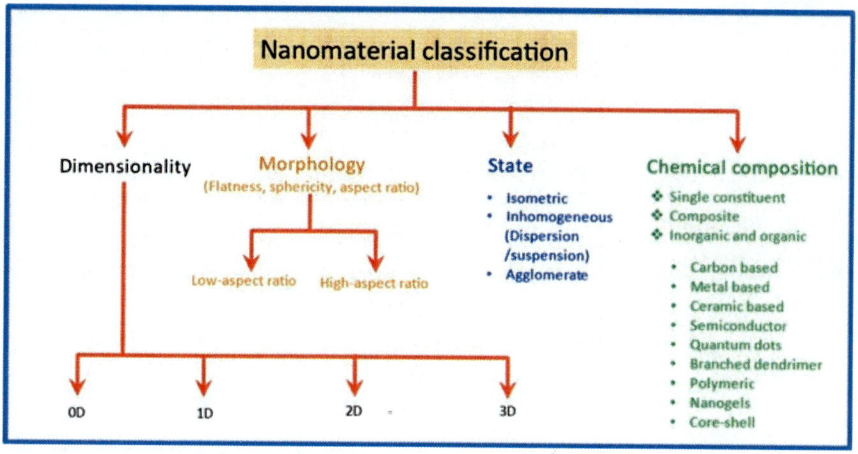

Fig. 1 Nanomaterials classification based on dimensionality, morphology, state, and chemical composition

whereas nanowire, nanohelices, nanozigzags, nanotubes, nanofibers, and nanobelts belong to high-aspect ratio nanomaterials.

Nanomaterials are also classified as single constituent nanoparticles and nanocomposites based on chemical composition. They can be subcategories as follows. Carbon-based materials, e.g., carbon nanotubes (CNTs), graphene, fullerenes, carbon nitrite, etc. Metal nanomaterials are synthesized using metals or metal derivatives like metal salts, organometallics, etc. Nanostructures having branch like structures are termed as branched dendrimers. Nanomaterials are combined with other materials as a matrix to improve toughness, mechanical strength, electrical, thermal conductivity, etc. Quantum dots (QDs) are small semiconductor particles where their particle size is in the range of 1–10 nm. The electronics and optics properties are different compared to bulk particles. They absorb UV light or white light and emit specific wavelength hence used in display applications (Halivni et al. 2015). Polymeric nanomaterials and lipid-based nanoparticles are extensively used in biomedical applications. Nanolipids act as vehicle for hydrophilic and hydrophobic molecules. It controls dug release at a site in the human body and has low toxicity.

1.2 *Methodologies and Characterization*

(a) Production Methods of Nanomaterials

Method employed for synthesis of nanomaterial governs the property. Hence, the right choice of synthesis method is important parameter that decides the property of nanomaterials. Nanomaterials are synthesized by bottom-up approach and top-down approach. In bottom-up approach (chemical method),

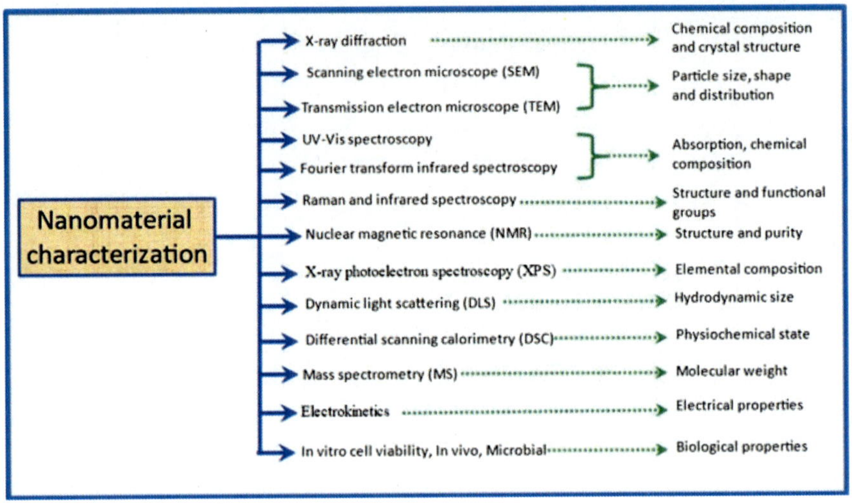

Fig. 2 Different techniques to characterize nanomaterials

synthesis will be started from atoms or molecules and then add one by one to form bigger particles, whereas in top-down (physical method), the synthesis will be started from bulk material, and then the particle size will be made smaller by different physical techniques. Bottom-up approach includes co-precipitation method, sol–gel synthesis, micro-emulsion method, hydro/solvo-thermal method, green synthesis, biochemical synthesis, chemical depositions like atomic layer deposition and vapor phase deposition, self-assembly, and DNA scaffolding of nanoelectronics (Burda et al. 2005; Patel et al. 2012, 2013). Ball milling, physical vapor deposition (PVD), and lithography-based techniques are examples of top-down approach (Burda et al. 2005; Velez et al. 2020).

(b) Characterization of nanomaterials

The synthesized nanomaterials with numerous application needs to be characterized for their different properties like, size, distribution, elemental composition, morphology, roughness, topography, optical, electrical, magnetic, mechanical, and thermal properties. Fig. 2 shows different characterization techniques.

1.3 Nanomaterials for Biomedical Applications

With the advancement of the technologies over the years, nanomaterial has been used in varied biomedical applications because of its versatility represents robust chemical, optical, electronic, and mechanical characteristics and believed to be keep on growing and expanding to overcome the limitation of biomaterials approaches

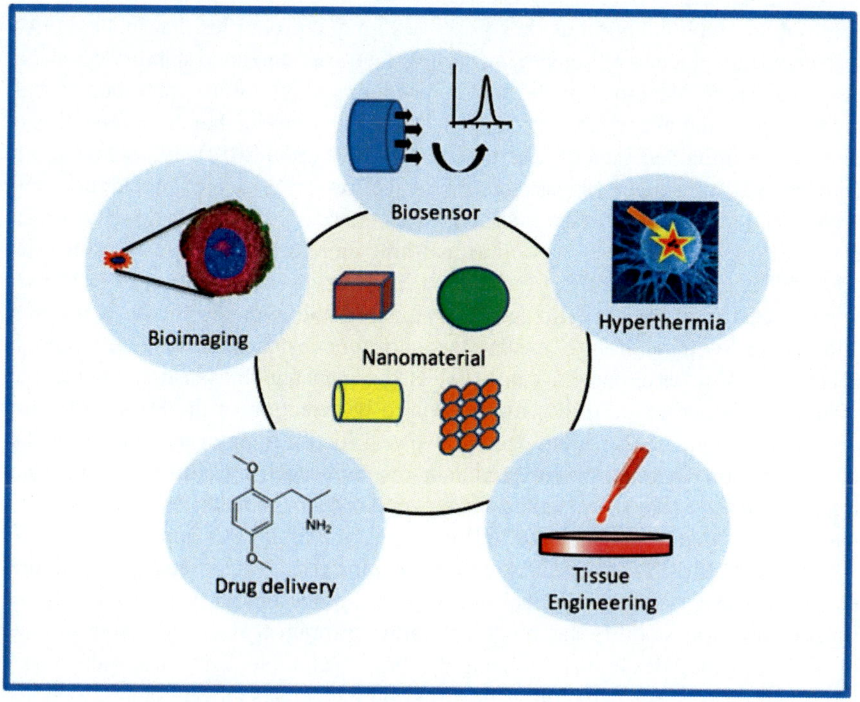

Fig. 3 Nanomaterials usage in biomedical applications

in conventional, clinical, and mechanic-based methods. The uses of nanomaterial have been applicable in major categories like diagnosis (bioimaging and biosensors), therapy (regenerative medicine, cancer treatment, and antibacterial agents), and drug delivery (gene delivery and medication delivery) as shown in Fig. 3.

1.3.1 Nanomaterial as Biosensors

Devices that interact with biological materials are known as biosensors. Typically, biosensors have three main components: (1) receptor or bioreceptor, (2) the transducer (signal generator), and (3) the electronic unit. The very first biosensor concept has been put forward by Clark and Lyons in (1962). Since then, biosensors have become the most studied in biomedical applications. Biosensors are classified on different measures that include type of transduction performed, like optical, electrochemical, or piezoelectric nature (Thévenot et al. 2001; Li and Lee 2020). The other classifications include enzymatic biosensors, immune sensors, apt sensors, and genosensors, (Kurbanoglu et al. 2017). The first and the foremost prerequisite of a biosensor device is to be fast, highly selective, and sensitive in terms of capturing the biochemical signals. The biosensors sensing abilities are directly proportional to the materials used (Perumal and Hashim 2014).To meet these requirements, nanomaterial received so much attention as they possesses unique properties that include physical, chemical, mechanical, electrical, high surface area, and good stability which has been extensively used so far in the field of development of biosensors. Nanomaterials are known for their easy fabrications and thus allow reproducible immobilization of bioreceptor units and increase the biocompatibility (Biju 2014). In comparison to conventional transducers, biosensors guarantee low cost, ease of preparation with high sensitivity. Biofilm-based microbial sensors are used to detect toxicity (Logan et al. 2019; Chu et al. 2020) as well as real-time monitoring (Prévoteau and Rabaey 2017; ElMekawy et al. 2018; Chu et al. 2020). Enzyme-based biosensors are used widely because of the catalytic power of enzymes and high specificity (Ronkainen et al. 2010; Rocha et al. 2021) and detect harmful substances in food (Kurbanoglu et al. 2020). Viruses or bacteria can be detected using a nanobiosensors; for example, reduced graphene oxide (rGO) film was used in detection of the H1N1 influenza virus (Joshi et al. 2020). In this section, we will focus on major nanomaterials that include carbon nanotubes, semiconductor quantum dots, hybrid (gold nanostructure, *Graphdiyne,* electrospunnano fibers, composite nanomaterials), and magnetic nanoparticles that have been used for biosensing and diagnostic applications.

Gold NPs (AuNPs) have been extensively explored in the nanosensing field (Doria et al. 2012) because of their optical and electronic properties, easy synthesis, surface functionalization, stability, and biocompatibility (Emrani et al. 2016). Shape and size of the gold nanoparticles can by using synthesis technique. Thus, it has been used in different forms, for e.g., gold nanowires and nanocubes (Karim et al. 2014) and as gold nanorods highly selective in a detection of DNA (Parab et al. 2010). Wang et al. (2015) reported super sandwich electrochemical DNA biosensor. Hosseini et al. developed a method for the detection of aflatoxin B1 (AFB1), based on the interactions of gold nanoparticles (AuNPs) with an aptamer (Hosseini et al. 2015). Aflatoxin B1 (AFB1) detection was developed by Hosseini et al., due to interactions between gold nanoparticles (AuNPs) and an aptamer (Hosseini et al. 2015). Due to their outstanding properties, AuNPs has become promising candidates for bioanalytics as well as for many other research applications. For example, cost-effective and simple enzyme-linked immunosorbent assay (ELISA) technique showed low sensitivity during diagnosis (Deng et al. 2019). Nanomaterials are used as enzymes to increase the sensitivity for different cancer types is a good strategy (Jiao et al. 2019; Gao et al. 2020).

Another example of nanomaterial being explored and utilized in bioanalytic applications is quantum dots (QDs). The major shortcoming in biosensor technology is its toxicity. To overcome this issue, QDs are coated using inert or biocompatible coatings (Biju 2014). The graphene-based QDs have excellent electrochemiluminescence (ECL); therefore, it has been utilized in the detection of biomacromolecules like nucleic acids, proteins, or glucose molecules (Xie et al. 2016; Kumawat et al. 2017). In near future, the QD-based biosensors will be an asset to biosensing applications.

Besides these, carbon-based nanomaterials (CBNs) are also the most promising nanomaterials in recent years which include carbon dots (CDs). The electrocatalytic property of CNDs has recently been reported in application of in electrochemical sensors toward oxygen reduction (Martínez-Periñán et al. 2018), exploited for glucose biosensing (Li et al. 2014) and DNA sensing (García-Mendiola et al. 2018).Carbon nanotubes (CNTs), another form of carbon-based nanomaterials, arranged hexagonally and give rise to single walled (SWCNTs) and multiwalled (MWCNTs) (Battigelli et al. 2013). The new cost-effective material-based carbon-based nanomaterials have been reinvented recently termed as carbon black (CB) having good electrical conductivity, solvents solubility, and fast electron kinetics (Mazzaracchio et al. 2019). It is widely used in different biosensor designs such as enzymes, DNA, and antibodies biosensors (Arduini et al. 2020). One study showed that CB as a substrate for enzyme immobilization in amperometric biosensors designing (Arduini et al. 2020). Nanodiamonds (NDs), new addition to the carbon nanoparticle, received much attention in biosensor applications (Zhang et al. 2011). For example, an electrochemical biosensor-containing copper, nanodiamond has been used for the detection of Parkia seeds (PS) amino acids (Babadi et al. 2019).Graphene (GR)-based nanoparticles are used in biosensor technology as the coating material of the sensor electrodes due to their inherent oxidation state, number of layers, and functional groups (Peña-Bahamonde et al. 2018).

Recently, magnetic nanoparticles turn out to be an alternative for fluorescent labels in biosensor devices. Moreover, they are more advantageous because of their size, physiochemical properties, and low cost. Iron oxide is the most common for bioanalytical applications among a wide range of ferromagnetic materials (Haun et al. 2010). Last but not least, novel nanocomposite material are being studied and showed their potential in the enhancement of the function of NPs in the field of biosensing applications like microfluidic biosensors (Singh et al. 2017). Bergdahl et al. have developed a capacitive sensor based on selective interaction between aminophenylboronic acid and diols immobilized on the surface of gold electrode (Bergdahl et al. 2019). Therefore, with the utilization of nanocomposites, biosensing platforms present substantial potential in improving testing diagnostics (Maduraiveeran and Jin 2020).

1.3.2 Nanomaterials in Medical Imaging and Contrast Enhancer

Medical imaging is the field of imaging at a cellular level for early stage diagnosis to understand the disease characteristics, its prognosis, and treatment efficacy evaluation thus absolutely important in terms of disease management. Magnetic resonance imaging (MRI), computed tomography (CT), positron emission tomography (PET), etc. are extensively used bioimaging techniques for diagnosis. The biggest advantage of these techniques is that they non-invasive and resolution of imaging is very high. However, sometimes, these imaging does not provide proper biological information. Thus, in vivo imaging depends on the agents that are generally used in bioimaging techniques for the identification of organ, tissue, or track cell migration called contrast agents that have an important role in the success of these techniques. For to overcome the toxicity, low retention time, and low imaging time associated with these contrast agents, different compounds have been identified as plausible contrasting agents (Berry and Curtis 2003; Rümenapp et al. 2012).Here comes the pivotal role of nanomaterials because of their adjusting property for high-performance and as an advanced molecular imaging agents (Nune et al. 2009). There are different nanomaterials that being studied showed promising results as magnetic resonance imaging contrast agents for imaging in cancer diagnostics to visualize tumors such as Fe_3O_4 NPs, QDs, Au NPs, and titanium dioxide (TiO_2). Moreover, formation of composite or hybrid nanomaterials makes them multifunctional in terms of different applications like imaging termed as "theranostic agents" (Ardelean et al. 2019).The iron oxide nanoparticles are used as contrast agents in CT and MRI imaging due tunable magnetism and high sensitivity for cancer cells (Chou et al. 2010). A few other nanomaterials that showed potential to be a good contrasting agent for the bioimaging techniques are luminescent semiconductor quantum dots (QDs) (Yao et al. 2018). Doped and undoped zinc sulfide are used as luminescent nanoprobes for biomedical imaging. The study by Yuan et al. showed that improved fluorescent probes of QD provides enhanced non-invasive tumor-targeting imaging system (Yuan et al. 2018). The conjugated ZnS-QDs i.e ZnS@CMC showed significant photoluminescence activity for bioimaging malignant glioma cells in vitro and thus possess potential applications in cancer research (Caires et al. 2020). Moreover, these QDs can be functionalized with surface modification, photostability, stable circulation time, and strong tumor specific homing property and thus can be used for multipurpose applications.

In recent years, titanium dioxide nanoparticles (TiO_2) emerged as promising agents. In conjunction with the fluorescent dye or magnetic resonance contrasting agents, these labeled TiO_2nanoparticles, nanotubes, or nanoprobes showed future prospects for cell imaging. It includes a few examples in which TiO_2 nanoparticles have been successfully prepared nanocontrast agent (nano-CA) as a candidate for molecular receptor targeted MR imaging (Chandran et al. 2011). In one study, the biocompatible mesoporoustitania nanoparticles (MTNs) were shown to have functionalized phosphate-containing molecules for intra-cellular bioimaging (Wu et al. 2011). In another study, TiO_2 with biocompatible glycol chitosan (GC)-coating coupled with gold nanoparticles showed excellent stability, biocompatibility for tumor-targeting CT imaging (Ryu et al. 2020).

Gold in form nanorods, nanocages, nanoshells and nanospheres has been used as stable contrast agent for bioimaging (Wang et al. 2005; Liao, et al. 2006; Choi et al. 2012). Cancer imaging is performed using gold nanorods by using several imaging tools (Huang et al. 2007; Oldenburg et al. 2009; Villar-Alvarez et al. 2018; Bai et al. 2020; Fernando et al. 2020). Smart capsule-based endoscopic cameras are used for in situ diagnostic and imaging. In future, these potential devices can be combined with nanobiosensor (Sim and Wong 2021).

1.3.3 Tissue Engineering

Tissue malfunctioning is one of the most noteworthy medical issues around the world. There are interdisciplinary scientific areas that are dealing with this collectively termed tissue engineering (TE) or regenerative medicine. TE is connecting link between engineering materials science, medicine, and biology. It was first introduced by Leland Kaiser in 1992 since then become the hottest research topic in medical science (Kaiser 1992). Tissue engineering is a field that deals with repairing, replacing, or regenerating human cells, tissues, or organs by using cells and/or a combination of cells with biologically active materials to restore or establish normal function as close to the body's native tissues. With the advancement of this field, current therapies got revolutionized and significantly improved the patient's life expectancy. The classic challenges encountered in TE are biocompatibility with the host, lack of functioning cells, the low mechanical strength of engineered cells, and nutrition limitation. To meet these challenges, there are many approaches are getting developed over a decade on searching for biomaterials that can be used as scaffolds. Among these, nanotechnology attracted scientists the most because of the nanometer size, surface area, appropriate physicochemical, mechanical properties, and biocompatibility for mimicking native tissues. In fact, the natural tissues and associated extracellular matrices are composed of nanostructured materials (LeDuc and Bellin 2006); therefore, it would be wise enough to put more efforts into designing and developing nanostructured materials.

Nanotechnology can be used to create nanofibres, nanopatterns, and controlled release nanoparticles. The nanomaterials can be the best choice as they are highly organized and present themselves as scaffolds. There are reported studies and examples where the nanomaterials have been a choice and proved to be an effective solution in TE, such as bone (Wen et al. 2012), ligament (Wang et al. 2007), vascular (Yin et al. 2013), heart valves (Iturri and Toca-Herrera 2017), plastic surgery (Biggs et al. 2017), joint substitutions (Dzedzickis et al. 2018), bladder tissue (He et al. 2010), and other restorative applications (Lin et al. 2019). The most common gold nanoparticles (GNPs) present themselves as an excellent candidate for bone tissue regeneration (Boisselier and Astruc 2009; Chen et al. 2015). GNPs also can be applied for wound healing applications (Akturk et al. 2016). Silver nanoparticles (AgNPs) are also widely used metallic NPs mainly for their antimicrobial properties for skin grafting and regeneration (Rai et al. 2009; Biswas et al. 2018). Titanium dioxide (TiO_2) nanoparticles are the most studied nanomaterial in TE especially in cardiac tissue engineering (Jawad et al. 2008) but also been used in bone tissue engineering as well. For skin TE, TiO_2 nanoparticles have been explored (Li et al. 2016).

Besides the metal-based, ceramic has received much importance for its usage as structural material because its stability under extreme conditions like temperature and chemical etching. For example, bioactive glass–ceramic NPs (nBGC) shown to have a better role in the development of bone tissue engineering over metal NPs and have been utilized for their excellent mechanical in bone tissue regeneration via improving scaffolds' performance (Vieira et al. 2017).

The polymeric-based nanoparticles (PNPs) are the most adjustable NPs for TE applications especially because of its degradation rate (Tang et al. 2016). Carbon nanotubes (CNTs) are the most commonly used polymeric-based nanomaterial in TE (Goncalves et al. 2016). Conductive polymers such as polyacetylene, PPy, polythiophene, polyfuran, and polyaniline (PANI) are commonly used in tissue engineering (Zare et al. 2019).

For the past few years, nanocomposites have been introduced in medical implants. Nanocomposites are types of materials formed by using two or more materials together in order to provide greater physiochemical properties (Breuer and Sundararaj 2004). There are examples like nanocomposites with titanium that showed efficient biocompatible medical implants (Fellah et al. 2019) like silver polytetrafluoroethylene (Ag-PTFE) nanocomposite (Wang et al. 2019) and NiTi nanocomposites (Razzaghi et al. 2020). Not only this, many of the nanomaterials present themselves as a promising candidate in the field of neural tissue regenerations like metallic NPs (Castro et al. 2018), silica NPs (Liu et al. 2020), magnetic NPs (Fang et al. 2019), quantum dots (Yin et al. 2020), and organic nanomaterials (Dos Santos et al. 2020).

1.3.4 Pharmaceutical Drug Delivery

Drug delivery platform includes approaches, formulations, manufacturing techniques, storage systems, and technologies for the transportation of the drug or medicine (pharmaceutical compound) to its target site for achieving desired medicinal effect. The first description of drug targeting and delivery was introduced in the early 1990s by Paul Ehrlich as "magic bullet paradigm" (Davis 1997). After that, this field gained so much attention because it involves targeted medication delivery as well as gene delivery for a varied range of biological health issues. Taking an example like the conventional treatment for cancers is chemotherapy which relies on the circulatory system to transport anticancerous drugs to the affected site. But it suffers serious side effects like non-specificity, drug toxicity which can harm nearby healthy cells and organs. Therefore, one needs to look for the technique which can direct the drug to the specific area avoiding side effects. In this respect, targeted medication stands out as an alternative approach. Natural drugs possess lower toxicity and side effects, low price, and good therapeutic potential (Mohanty et al. 2017). However, natural compounds lag behind, and still many of them are not clearing the clinical trials (Watkins et al. 2015). Thus, researchers prefer to explore the available chemical libraries to discover novel drugs. Site-specific delivery, in vivo instability, poor absorption in the body, poor solubility, poor bioavailability, and programmed release of the drugs due to physiological targets have been the major issues associated with the molecular and macromolecular (large size materials) therapeutic approaches (Jahangirian et al. 2017). Therefore, for sustained drug delivery, development of advanced technology is of utmost importance for the researchers. Hence, nanotechnology opens up the opportunities in advanced drug formulation, in altering the property of the materials, in controlling drug delivery and release successfully. Several research studies have proven that nanosized materials could be utilized as drug carrier by encapsulation of drugs to target tissues precisely (Jahangirian et al. 2017). Moreover, because of its small size, it can move freely in the human body tends to be taken up by different cells inside the body instead of getting accumulated in the spleen and liver (Rawal and Patel 2019). Nanoparticles are suitable for the broad range from cells to tissue to organs and can be achieved by their highly tunable size, shape, and surface functionalization properties to target a specific tissue.

Therefore nano-based drug delivery systems (DDS) are now being extensively studied so as to facilitate efficient drug delivery. Nanocarriers have the parameters like different compositions (e.g., organic, inorganic, and hybrid materials) and have the ability to attain different forms when associated drugs (core–shell system or matrix system) for understanding the drug delivery profile (Siepmann et al. 2008). The very first application of nano-based DDS was lipid vesicle most popularly known as liposomes and micelles which are now FDA approved (Bozzuto and Molinari 2015). Liposomes are of spherical form composed of phospholipids and steroids which are analogous to the cell membranes therefore considered as a better drug delivery vehicle (Bozzuto and Molinari 2015). It can deliver a range of bioactive compounds like antioxidants and proteins (Reimhult 2015; Simão et al. 2015). Since last two decades, lipid-based NPs have been widely employed as drug delivery systems for diseases, gene therapy, antiviral vaccines, and among other nanomedicine uses (Fenton et al. 2018; Anselmo and Mitragotri 2019) due to its uptake ability by reticuloendothelial system (Alyautdin et al. 2014). The nano-based DDS requirement includes surface properties, particle size; solubility, increase permeation, easily modified or functionalized and controlled release of drugs at a constant rate and time (Hossen et al. 2019).The nanomaterials meet all these requirements and thus provide advantages for pharmaceutical industry to improve patients' life quality (Rizvi and Saleh 2018). The nanomaterials which are getting utilized besides liposomes are carbon-based NPs, metal nanoparticles, inorganic NPs, QuantomDotss, polymeric NPs, and their nanocomposites for better and more efficient DDS.

Metal nanoparticle, silver and gold have the surface plasmon resonance (SPR) property which was lacking in liposomes or micelles and have better biocompatibility (Kudr et al. 2017). Multiple therapeutic agents or large biomolecules like antibiotics, proteins, genes (DNA and RNA), and varied ligands can be easily conjugated or (bio) functionalized to AuNPs' surface and thus deliver the drug with the controlled release through biological stimuli or light activation (Kong et al. 2017). Very few studies have been done with silver nanoparticles as drug delivery agent (Prusty and Swain 2018). There are other kinds of metallic nanoparticles, such as zinc, titanium, platinum, selenium, gadolinium, palladium, cerium dioxide which are also getting exploited for the biomedical applications (McNamara and Tofail 2017).

Carbon-based NPs like graphene is also one of the candidate for drug delivery (Li et al. 2017). Among the different carbon allotropes, carbon nanotubes (CNTs) have been explored as a highly competent drug delivery agent (Faraji and Wipf 2009) because of its non-invasive administration (Panczyk et al. 2016). CNTs represent one of the most investigated families of nanocarriers for cancer therapy (Chen et al. 2017). Like other NPs, the CNTs also have drawbacks in terms of their sustained release inside the cell and its stability. But it can achieved by utilizing inner hollow cavity of CNT that can isolate the drug from the outer environment (Maiti et al. 2019).

The coupling of nanoscience with natural products is very attractive and rapidly growing in recent times due to their functional properties, such as inducing tumor-suppressing autophagy and as antimicrobial agents. The group of materials is termed polymer nanocomposites (PNCs). There are numerous biopolymeric materials that have been utilized in the nanofabrication of nanoparticles because of their high biocompatibility and biodegradability properties (Watkins et al. 2015) and thus being extensively used in DDS. Chitosan is a natural polymer utilized for wide range of drug release systems of various types, including buccal, intestinal, eye, and pulmonary (Artursson et al. 1994; De Campos et al. 2001; Portero et al. 2002; Al-Qadi et al. 2012). Alginate is another natural biopolymeric. Patil and Devarajan developed insulin-containing alginate nanoparticles with nicotinamide as a permeation agent (Patil and Devarajan 2016). Cellulose and its derivatives also resulted in the controlled release of drugs (Sun et al. 2019).

In a nutshell, several studies conducted regarding release mechanisms of drugs in nanocarriers showed that these nanostructures can stay in the blood circulatory system for a longer period of time and enable the controlled release of drugs at the targeted location. But still, there are challenges that need to be taken into account for the successful delivery of drugs to their target sites. Developing smart and functionalized nanosized drug carriers with their great properties can contribute to the enhancement of human health lifestyle.

1.3.5 Hyperthermia

Cancer is the most dreaded disease that is responsible for most of mortality worldwide. There are conventional methods like surgery, radiation therapy, chemotherapy, and gene and immunotherapy available for treating cancers but still suffer failure due to several limitations. With the advancement of technologies, there is still an urgent need for efficient methods for cancer therapy. In oncology, the term hyperthermia (HT) describes the therapeutic treatment in which heat is applied using an external energy source or increasing the temperature of the infected/diseased area to 41–46°C to destroy cancerous/tumor cells and tissues without affecting healthy cells/tissues. HT can be an adjuvant therapy when coupled with the available conventional therapies can work synergistically in cancer therapy treatments. There are reports available on certain tumors where HT has been applied in conjunction with radiotherapy or chemotherapy significantly improves tumor control, complete response, and survival rates of the patients (Roussakow 2018). HT treatment effectiveness is greatly influenced by temperature profile at the targeted tumor site (Wust et al. 2002).Therefore, different types of HT developed, i.e., local (small area tumor), regional (large tissue), and whole-body hyperthermia (metastatic cancer). Traditionally HT treatment is performed using different energy sources like microwave, radiofrequency, laser, ultrasound, and tubes with hot water to heat the tumor. Despite using different techniques, HT has not yet been established well in clinical routine due to several limitations. These include difficulty in achieving uniform heat dispersion throughout the tumor, failure to heat the target without damaging the nearby healthy cells, thermal under dosage in the target region, and dissipation of heat by the blood which can lead to serious side effects (Jordan 2006). So to address these drawbacks, new efficient HT methods are being developed that led to the application of nanomaterials as hyperthermia agents.

Among nanoparticles, magnetic nanoparticles (MNPs) are the first one to be explored in the HT treatment for cancer in 1957 (Gilchrist et al. 1957) ever since then new approaches have been evolved and will be utilized in future cancer therapies. Because MNPs possess unique features like heating mediator; tumor focused, uniform heat dissipation, energy absorption, minimally invasive, and biocompatibility make them a promising candidates for HT (Ito et al. 2005; Mornet et al. 2006). These MNPs have been applicable in all HT techniques like for local HT. Hyperthermia using MNPs has developed a new therapeutic system termed as magnetic nanoparticle hyperthermia therapy (MHT) for ablation of malignant tumors (Tian et al. 2015). MHT is promising because it can induce hyperthermia in tumors at any body's location and is therefore helpful in treating poorly accessible and deep-seated tumors. Moreover, these MNPs accumulate in the tumor thus limiting healthy tissue damage (Bañobre-López et al. 2013) followed by shrinkage of the tumor. MHT involves several types of iron oxides for the synthesis of MNPs, like magnetite (Fe_3O_4), hematite (α-Fe_2O_3), and maghemite (γ-Fe_2O_3) (Faraji et al. 2010; Bañobre-López et al. 2013). Among these, superparamagnetic iron oxide nanoparticles (SPIONs) exhibits excellent properties that enable uniform intratumoral distribution and controlled heating of tumors without having side effects (Jordan et al. 2006). The first report by Shinkai and group showed the effectiveness of MHT in gliobalstoma (GBM) (Shinkai et al. 1996). Thereafter, several studies have shown the potential of MHT in cancer therapy alone as well as in combination with other therapies. Besides GBM, MHT has been explored in treatment of other cancers as well like prostrate, esophagus, and liver cancers (Johannsen et al. 2007; Kunjachan et al. 2015). Moreover, MHT is one of the first applications of nanotechnology in biomedicine to be tested in clinics. Clinical studies of MHT have been introduced in 2001 by Jordan et al. (2001). Afterward, several clinical trials were performed to explore the potential role of MHT in treatment for varied cancers like prostrate, esophagus, liver, and GBM (Johannsen et al. 2007; Maier-Hauff et al. 2011; Kunjachan et al. 2015).

Gold nanoparticles have also been extensively used in hyperthermia treatment. Because of its high atomic number, gold shows a synergist effect with radio therapy upon induction by laser light. These AuNPs have been utilized in a method defined as nano-photo thermal therapy (NPTT) or photodynamic therapy (PDT) in which selective killing of tumor-induced by specific targeting of nano-photosensitizer such as gold nanoparticles (AuNPs), into the cancerous tissue when induced by light. These AuNPs change the photothermal properties of the medium when exposed to laser and thus induced localized hyperthermia (Burda et al. 2005; Schwartz et al. 2011). Different forms of AuNPs like gold nanoshell (AuNS), gold nanorod (AuNR), and spherical AuNPs have different photothermal responses (Pattani and Tunnell 2012). GNRs have been conjugated with molecules like folic acid to target and destroy cancerous cells specifically (Mehdizadeh et al. 2014).

Carbon-based nanomaterials like carbon nanotubes (CNTs) (Zhou et al. 2009), fullerene (Chen et al. 2012), and graphene (Yang et al. 2013) have recently explored and showed great potential as NPTT agents. Graphene and its derivatives showed as promising NPTT agents (Yang et al. 2013). CNTs-mediated HT is a new field in comparison to MNP and GNR-driven hyperthermia. The CNTs have opened new ways for the novel nanostructures development due to their ability to engineer surface conjugation with varied molecules and their unique physiochemical properties for anticancer hyperthermia therapy (Iijima 1991; Fabbro et al. 2012). So keeping in mind all together, the properties imparted by NPs like toxicity, biocompatibility, and heating efficiency can facilitate thermo-chemo-radio therapy for the anticancer treatments.

1.4 Toxicology of Nanomaterials

There are very good review articles in the literature explaining the toxicology of nanomaterial affecting the human and the surrounding environment (Ganguly et al. 2018; Peng et al. 2020). Here in this chapter, we are going to focus on the aspects of nanomaterial toxicity in the biomedical field. As we all get to know now nanomaterial is of great importance in research and development especially in biological applications like drug delivery, biomedicine, and anticancer therapeutics. A single variation or modification in the nanomaterial can cause a drastic change in their properties thus can impact living cells or the human body. Therefore, despite the promising biomedical application of nanomaterials, one cannot ignore the side effects or toxicological effects caused by these engineered or modified nanomaterials. So establishing their effects and fate inside the human body is a challenging task. One must perform a detailed toxicity assessment study in order to under the potential toxicity caused by these materials. Therefore, a new branch of toxicology has been developed termed nanotoxicology that deals with the knowledge of toxicity caused by nanomaterials or nanoparticles on living organisms. It includes the basic understanding of the possible physiochemical properties of nanomaterials governing the adverse effects or cytotoxicity, physiological impact on cells, uptake pathways, and plausible ways of interaction or human exposure to the nanomaterial.

If we discuss in detail, then this branch deals with as follows:

(1) The induced toxicity depends on different physicochemical properties of nanomaterials including the nanoparticles size, shape, morphology, aspect ratio, surface coating, and charge.
(2) The means or routes of exposure by which these nanomaterials can cause serious health issues include ingestion, inhalation (respiratory tract), skin contact, gastrointestinal tract, crossing the blood–brain barrier, accumulating in the liver and spleen.
(3) The mechanism by which it causes toxicity like reactive oxygen species generation and results in oxidative stress, inflammation, cytotoxicity, genotoxicity, and other antioxidant pathways.

2 Nano Toxicity Based on Physiochemical Properties of Nanomaterials

The nanotoxicity depends upon nanomaterials size, shape, aspect ratio, and surface functionalization (Oberdörster 2010). It was stated that nanoparticle is inversely proportional to the toxicological effects as the size decreases, the surface area/volumeratio increases exponentially that can lead to the increment of biological and chemical reactions (Johnston et al. 2010). Moreover, NPs having the same chemical composition can exhibit different toxicities based on their surface area and size. For example, the SWCNT and MWCNT—the two different states of carbon nanotube delivered different toxicity (Fraczek et al. 2008; Di Giorgio et al. 2011).There are many studies have been conducted on CNTs toxicity (Kobayashi et al. 2017). While the smaller size of graphene quantum dots (GQDs) possesses high cell viability over graphene oxide (GO) (Wang et al. 2016). There are other studies have been reported the size dependent toxicity like Cho et al. showed acute toxicity in BALB/c mice after injecting the AgNPs of different size (10, 60 and 100 nm) (Cho et al. 2018). Lopez et al. evaluated the toxic effects of three different sizes of AuNPs (10, 30 and 60 nm) and showed that the AuNPs of 10 and 30 nm crossed the nuclear membrane resulting into DNA breakage while 60 nm AuNP accumulated in the spleen (Lopez-Chaves et al. 2018).

The second factor that influences the toxicity imparted by nanomaterials is their shape. The typical shapes of NPs are rod, spherical, sheet, and cylinder which plays a crucial role in elucidating their biological as well as toxic activity. It has been reported that needle-shaped mesoporous silica NPs exhibit more toxicity (Huang et al. 2011; Li et al. 2013) over spherical shape. Similarly, silver nanoplates were found to be more harmful than that of silver nanospheres (Abramenko et al. 2018). Different shapes of AuNPs including stars shaped, rods, as well as spheres have been examined for their cytotoxicity against varied human cancer cell lines (Steckiewicz et al. 2019). Kang et al. found that elongated-shaped MWCNTs were more cytotoxic than spherical-shaped carbon NPs and led to a considerable elevation in intracellular ROS levels compared with spherical-shaped carbon NPs (Kang et al. 2015).

An aspect ratio (length/width) of the nanoparticle also plays a crucial role in defining their toxic behavior, greater the ratio higher the toxicity (Lippmann 1990). A subclass of nanoparticles that considered high-aspect ratio nanoparticles is nanotubes, nanorods, and nanowires (Oberdörster et al. 2007). Recent study by Abtahi et al. showed the implication of aspect ratio on the uptake and nanotoxicity of AuNPs and showed increasing the size and aspect ratio enhanced internalization and filtered out at a higher rate (Abtahi et al. 2019).

Another concept that is getting introduced into the toxicology assessment is the surface functionalization of NPs. Conventionally inside the cellular environment, NPs interact with body fluids (media of proteins and biomolecules) before making contact with the cell. The protein that tends to get absorbed on the surface of the NP forms a NP-protein complex known as the corona (Pederzoli et al. 2017). Therefore, careful studies need to be conducted based on corona systems to decide its ultimate biological fate. The surface modification/functionalization or coatings are applied in order to modify the properties of NPs like their stability, wettability (Albanese et al. 2012; Egbuna et al. 2021). Zhou group showed that modified graphene sheet can overcome the toxic effect of graphene oxide (Chong et al. 2015). On the contrary, there are studies where surface modification tends to enhance the toxicity level of the NPs. For example, it has been seen that positively charged NPs internalized more efficiently through negatively charged lipid bilayer membrane and result in more toxicity than neutral or negatively charged particles (Pozzi et al. 2014).The type and degree of surface modification or functionalization of CNTs showed CNT-mediated toxicity in different studies. Jiang et al. demonstrated that surface functionalization of MWCNTs with polyethylene glycol (PEG) reduces cellular uptake, intracellular ROS generation (Jiang et al. 2013).

3 Nanoparticle-Induced Mechanism that Causes Toxicity

Oxidative stress: This is the most common response exhibited by cells upon induction by nanoparticles. It is defined as a disturbed balance between antioxidants activities and the generation of oxidants, i.e., reactive oxygen species (ROS) that leads to protein lipid and nucleic acids oxidation (Sies 1991). Subsequently, the accumulation of these oxidized products disrupts the homeostatic balance that can be deleterious to the cells or organisms eventually leading to genotoxicity, cytotoxicity, and various diseases (Egbuna and Ifemeje 2017; Ajdary et al. 2018; Morsy et al. 2021). It can cause damage to DNA in different manners like DNA strand breakage and aberrant genetic mutations (Hengstler 2003; Saud Alarifi et al. 2013). The NP-induced oxidative stress can be categorized into direct or indirect oxidative stress. Direct oxidative stress involves direct induction of stress caused by ROS generation on the surface of NP via different mechanisms such as photocatalysis activity of TiO_2 nanoparticles and chemical reactions of metal ions released from nano-objects (Gurr et al. 2005). Whereas indirect involves ROS generation caused by mitochondrial dysfunction (Nichols et al. 2018). There are reports on metallic NPs causes oxidative stress and triggers different pathways available in the literature. Hou et al. reported that the zinc oxide NP is responsible for failure in maintaining the mini chromosome in the cell cycle pathway (Hou et al. 2019). The release of metal ions is also an important factor responsible for oxidative stress. For example, zinc oxide NP showed toxicity because of the release of Zn^{2+} ions thus increasing intracellular ROS levels in mouse macrophage cells (Song et al. 2010). Silver NP (AgNP) despite of having excellent antimicrobial activity is known to have cytotoxic properties due to oxidative stress (Kovvuru et al. 2015). AgNPs have also been shown to affect the cell cycle and apoptosis via ROS generation and mitochondrial dysfunction in HepG2 and rat tracheal epithelial cells (Xue et al. 2016; Tang et al. 2019). Likewise, AuNPs which has been widely used for anticancer therapeutics showed oxidative stress-mediated cytotoxicity on several cell (Mateo et al. 2014; Enea et al. 2020). The TiO_2 nanoparticles have the ability in inducing intracellular ROS levels but depend on the type of protein adsorption capability. For instance, in one study, it was shown that TiO_2 NPs induce intracellular ROS levels without photo activation (Horie et al. 2009), while in another study, it was showed that coating of TiO_2 NPs with alumina or silica prevents induction of intracellular ROS (Horie et al. 2010). Besides metal-based NPs, carbon-based NP majorly carbon nanotubes CNTs are shown to have responsible for toxicity through oxidative stress (Alshehri et al. 2016).There are reports which showed that both SWCNTs and MWCNTs increase the production of ROS in various types of cells, such as macrophages, fibroblasts, and epithelial cells (Dong and Ma 2015). Moreover, these CNTs were also responsible for the activation of cellular signaling pathways due to ROS (Jović et al. 2020).

Genotoxicity: We have already cited several examples above where oxidative stress eventually leads to cytotoxicity and genotoxicity. Here just to have a clear picture, genotoxicity is a type of toxicity that causes the destruction of the genetic material of the cell like mutation of the DNA strands (Cao et al. 2017). The different studies conducted using AgNPs in different organisms showed DNA damage, a nuclear aberration in them due to AgNPs (Krishnaraj et al. 2016; Pandiarajan and Balaji 2018).

Neurotoxicity: The brain is very much vulnerable to the changes especially oxidative stress and low antioxidant defense mechanism along with an abundance of lipids and proteins (Cui et al. 2004). A varied range of nanomaterials that includes polymeric nanoparticles, dendrimers, metallic nanoparticles, carbon nanotubes, and QDs have been developed and utilized for brain disease therapy (Shafiee and Kargar 2016; Huang et al. 2017). However, neurotoxicity associated with each class of NPs also reported. It includes study cisplatin-containing liposome NP associated with neurotoxic effect (Huo et al. 2012). Dendrimers, one of the widely used NP in brain therapy, known to induce neurotoxicity (Hammer et al. 2017).In case of silver NPs, release of silver ions cause damage to cell membrane integrity thus leads to cell necrosis (Sun et al. 2016). It has been documented that CNTs also have the capability to enter the brain through olfactory administration leads to their accumulation eventually causing inflammation of microglial cells (Mahmoudi et al. 2011). Studies have also showed that MWCNTs is responsible for higher neurotoxicity as compare to SWCNTs (Gholamine et al. 2017). Besides that, the longer CNTs in higher concentration are more neurotoxic as compare to their counterparts (Shi et al. 2017). The quantum dots also impart neurotoxicity based on their physicochemical as well as surface coatings and purity (Wu et al. 2016).

However, there are some reports stated that these nanoparticles do not always induce oxidative stress. For example, superparamagnetic iron oxide nanoparticles did not induce cellular oxidative stress in oligodendrocyte cells (Sruthi et al. 2018). Some studies have reported that cerium oxide (CeO_2) nanoparticles induce oxidative stress in culture cells (Ali et al. 2015), while some suggested that CeO_2 nanoparticles reduce cellular oxidative stress by acting as an antioxidant (Azari et al. 2019; Carvajal et al. 2019). Therefore, a careful interpretation of the relationship between toxicity and physicochemical properties of NPs is very important so as to minimize false results (Kroll et al. 2009). Moreover nanotoxicity is the second major field about the ultimate biological fate of nanoparticle toxicity thus needed precise instrumentation, medium conditions, and reliable *in cellulo* assays.

4 Effect of Nanotoxicity on Various Systems

The existing toxicological research on nanomaterials showed that NPs may have a high risk of toxicity than larger particles (Hoet et al. 2004). These NPs have proven to be toxic to humans, animals and environment as well (Oberdörster et al. 2005). Here in this section, we will discuss nanotoxicity's impact in affecting human systems like diseases asthma, dermatitis, lung disease, lung contaminations, tuberculosis, immune system illnesses, and so on. The first line of infection caused by nanomaterials is through inhalation ingestion and skin contact, if the skin is damaged these can easily access the bloodstream. Though the target distribution sites of the NPs are not known well, it appears that the liver and spleen are the main target organs (Hussain et al. 2005; Handy et al. 2008). It basically affects the reticuloendothelial systems (RES) in the liver after getting absorbed from the gastrointestinal tract (Manke et al. 2013). Moreover, there is hepatocyte malfunction caused by NPs. Studies showed that the Fe_3O_4 NP-induced loss of mitochondrial membrane potential in human hepatoma cells. The other NPs like fine carbon black, nickel, and TiO_2 particles were also shown to have an enhanced lung inflammatory response (Grassian et al. 2007; Pettibone et al. 2008). Cho et al. showed a massive pulmonary inflammatory response in rats infused with Co_3O_4- NPs (Cho et al. 2012). The NPs toxicity studies documented its effect on the cardiovascular system also for ex. exposure to single-walled carbon nanotubes and ultrafine carbon black resulted in cardiovascular effects (Li et al. 2007; Simeonova and Erdely 2009).

5 Conclusions and Future Prospects

A wide variety of nanomaterials with different dimensions and chemical composition play a pivotal role in determining the property. The right synthesis methodology is used to get desired characteristics from these nanomaterials. Synthesis conditions like temperature, solvent, and surfactant are important to obtained characteristic nanomaterial toward targeted applications. Co-precipitation, hydrothermal, and sol–gel synthesis are widely used for synthesis. Advancement in nanotechnology resulted in improved diagnosis, treatment of human disease. But nanoparticles with smaller size, high-aspect ratio, chemical composition, or surface modification result in an increase in toxicity. To overcome some of the adverse effects, green synthesis is employed recently for nanomaterial synthesis using plant extracts which results in lower toxicity. Nanomaterials with zero or minimal toxicity need to be developed. Researchers and scientists are interpreting cautiously the relation between toxicity and physicochemical properties to minimize false results.

References

N.B. Abramenko, T.B. Demidova et al., Ecotoxicity of different-shaped silver nanoparticles: case of zebrafish embryos. J. Hazard. Mater. **347**, 89–94 (2018)

S. Abtahi, R. Trevisan et al., Implications of aspect ratio on the uptake and nanotoxicity of gold nanomaterials. NanoImpact **14**, 100153 (2019)

M. Ajdary, M.A. Moosavi et al., Health concerns of various nanoparticles: a review of their in vitro and in vivo toxicity. Nanomaterials **8**(9), 634 (2018)

O. Akturk, K. Kismet et al., Wet electrospun silk fibroin/gold nanoparticle 3D matrices for wound healing applications. RSC Adv. **6**(16), 13234–13250 (2016)

S. Al-Qadi, A. Grenha et al., Microencapsulated chitosan nanoparticles for pulmonary protein delivery: in vivo evaluation of insulin-loaded formulations. J. Control. Release **157**(3), 383–390 (2012)

A. Albanese, P.S. Tang et al., The effect of nanoparticle size, shape, and surface chemistry on biological systems. Annu. Rev. Biomed. Eng. **14**, 1–16 (2012)

D. Ali, S. Alarifi et al., Cerium oxide nanoparticles induce oxidative stress and genotoxicity in human skin melanoma cells. Cell Biochem. Biophys. **71**(3), 1643–1651 (2015)

R. Alshehri, A.M. Ilyas et al., Carbon nanotubes in biomedical applications: factors, mechanisms, and remedies of toxicity: miniperspective. J. Med. Chem. **59**(18), 8149–8167 (2016)

R. Alyautdin, I. Khalin et al., Nanoscale drug delivery systems and the blood–brain barrier. Int. J. Nanomed. **9**, 795 (2014)

A.C. Anselmo, S. Mitragotri, Nanoparticles in the clinic: an update. Bioeng. Trans. Med. **4**(3), e10143 (2019)

I.L. Ardelean, D. Ficai, et al., Hybrid magnetic nanostructures for cancer diagnosis and therapy. Anti-Cancer Agents Med. Chem. (Formerly Current Medicinal Chemistry-Anti-Cancer Agents) **19** (1), 6–16 (2019)

F. Arduini, S. Cinti et al., Carbon black as an outstanding and affordable nanomaterial for electrochemical (bio) sensor design. Biosens. Bioelectron. **156**, 112033 (2020)

P. Artursson, T. Lindmark et al., Effect of chitosan on the permeability of monolayers of intestinal epithelial cells (Caco-2). Pharm. Res. **11**(9), 1358–1361 (1994)

A. Azari, M. Shokrzadeh et al., Cerium oxide nanoparticles protects against acrylamide induced toxicity in HepG2 cells through modulation of oxidative stress. Drug Chem. Toxicol. **42**(1), 54–59 (2019)

F.E. Babadi, S. Hosseini et al., Electrochemical investigation of amino acids Parkia seeds using the composite electrode based on copper/carbon nanotube/nanodiamond. J. Environ. Chem. Eng. **7**(2), 102979 (2019)

X. Bai, Y. Wang et al., The basic properties of gold nanoparticles and their applications in tumor diagnosis and treatment. Int. J. Mol. Sci. **21**(7), 2480 (2020)

M. Bañobre-López, A. Teijeiro et al., Magnetic nanoparticle-based hyperthermia for cancer treatment. Rep. Pract. Oncol. Radiother. **18**(6), 397–400 (2013)

A. Battigelli, C. Ménard-Moyon et al., Endowing carbon nanotubes with biological and biomedical properties by chemical modifications. Adv. Drug Deliv. Rev. **65**(15), 1899–1920 (2013)

G.E. Bergdahl, M. Hedström et al., Capacitive saccharide sensor based on immobilized phenylboronic acid with diol specificity. Appl. Biochem. Biotechnol. **188**(1), 124–137 (2019)

C.C. Berry, A.S. Curtis, Functionalisation of magnetic nanoparticles for applications in biomedicine. J. Phys. D Appl. Phys. **36**(13), R198 (2003)

M. Biggs, D. Zeugolis, et al., Foreword to special issue on two-dimensional biomaterials in regenerative medicine. Nanomed.: Nanotechnol. Bio. Med. **14** (7), 2351–2353 (2017)

V. Biju, Chemical modifications and bioconjugate reactions of nanomaterials for sensing, imaging, drug delivery and therapy. Chem. Soc. Rev. **43**(3), 744–764 (2014)

D.P. Biswas, N.M. O'Brien-Simpson et al., Comparative study of novel in situ decorated porous chitosan-selenium scaffolds and porous chitosan-silver scaffolds towards antimicrobial wound dressing application. J. Colloid Interface Sci. **515**, 78–91 (2018)

E. Boisselier, D. Astruc, Gold nanoparticles in nanomedicine: preparations, imaging, diagnostics, therapies and toxicity. Chem. Soc. Rev. **38**(6), 1759–1782 (2009)
G. Bozzuto, A. Molinari, Liposomes as nanomedical devices. Int. J. Nanomed. **10**, 975 (2015)
O. Breuer, U. Sundararaj, Big returns from small fibers: a review of polymer/carbon nanotube composites. Polym. Compos. **25**(6), 630–645 (2004)
C. Burda, X. Chen et al., Chemistry and properties of nanocrystals of different shapes. Chem. Rev. **105**(4), 1025–1102 (2005)
A.J. Caires, A.A. Mansur et al., Green synthesis of ZnS quantum dot/biopolymer photoluminescent nanoprobes for bioimaging brain cancer cells. Mater. Chem. Phys. **244**, 122716 (2020)
Y. Cao, Y. Gong et al., The use of human umbilical vein endothelial cells (HUVECs) as an in vitro model to assess the toxicity of nanoparticles to endothelium: a review. J. Appl. Toxicol. **37**(12), 1359–1369 (2017)
S. Carvajal, M. Perramón et al., Cerium oxide nanoparticles protect against oxidant injury and interfere with oxidative mediated kinase signaling in human-derived hepatocytes. Int. J. Mol. Sci. **20**(23), 5959 (2019)
A.G. Castro, M. Diba et al., Development of a PCL-silica nanoparticles composite membrane for guided bone regeneration. Mater. Sci. Eng., C **85**, 154–161 (2018)
P. Chandran, A. Sasidharan et al., Highly biocompatible TiO 2: Gd 3+ nano-contrast agent with enhanced longitudinal relaxivity for targeted cancer imaging. Nanoscale **3**(10), 4150–4161 (2011)
Y. Chen, N. Li et al., A dual targeting cyclodextrin/gold nanoparticle conjugate as a scaffold for solubilization and delivery of paclitaxel. RSC Adv. **5**(12), 8938–8941 (2015)
Z. Chen, L. Ma et al., Applications of functionalized fullerenes in tumor theranostics. Theranostics **2**(3), 238 (2012)
Z. Chen, A. Zhang, et al., The advances of carbon nanotubes in cancer diagnostics and therapeutics. J. Nanomater. (2017)
S. Chikazumi, S. Taketomi et al., Physics of magnetic fluids. J. Magn. Magn. Mater. **65**(2–3), 245–251 (1987)
W.-S. Cho, R. Duffin et al., NiO and Co, 3O4 nanoparticles induce lung DTH-like responses and alveolar lipoproteinosis. Eur. Respir. J. **39**(3), 546–557 (2012)
Y.-M. Cho, Y. Mizuta et al., Size-dependent acute toxicity of silver nanoparticles in mice. J. Toxicol. Pathol. **31**(1), 73–80 (2018)
W.I. Choi, A. Sahu et al., Photothermal cancer therapy and imaging based on gold nanorods. Ann. Biomed. Eng. **40**(2), 534–546 (2012)
Y. Chong, C. Ge et al., Reduced cytotoxicity of graphene nanosheets mediated by blood-protein coating. ACS Nano **9**(6), 5713–5724 (2015)
S.-W. Chou, Y.-H. Shau et al., In vitro and in vivo studies of FePt nanoparticles for dual modal CT/MRI molecular imaging. J. Am. Chem. Soc. **132**(38), 13270–13278 (2010)
N. Chu, Q. Liang et al., Microbial electrochemical platform for the production of renewable fuels and chemicals. Biosens. Bioelectron. **150**, 111922 (2020)
L. Clark, C. Lyons, Biosensors-a practical approach. Jnr. Ann. NY Acad. Sci **102**, 29–45 (1962)
K. Cui, X. Luo et al., Role of oxidative stress in neurodegeneration: recent developments in assay methods for oxidative stress and nutraceutical antioxidants. Prog. Neuropsychopharmacol. Biol. Psychiatry **28**(5), 771–799 (2004)
S. Davis, Biomedical applications of nanotechnology—implications for drug targeting and gene therapy. Trends. Biotechnol. **15**(6), 217–224 (1997)
A.M. De Campos, A. Sanchez et al., Chitosan nanoparticles: a new vehicle for the improvement of the delivery of drugs to the ocular surface. Application to cyclosporin A. Int. J. Pharm. **224**(1–2), 159–168 (2001)
D. Deng, Y. Hao et al., A colorimetric enzyme-linked immunosorbent assay with CuO nanoparticles as signal labels based on the growth of gold nanoparticles in situ. Nanomaterials **9**(1), 4 (2019)
M.L. Di Giorgio, S. Di Bucchianico et al., Effects of single and multi walled carbon nanotubes on macrophages: cyto and genotoxicity and electron microscopy. Mutat. Res./genet. Toxicol. Environ. Mutagen. **722**(1), 20–31 (2011)

J. Dong, Q. Ma, Advances in mechanisms and signaling pathways of carbon nanotube toxicity. Nanotoxicology **9**(5), 658–676 (2015)

G. Doria, J. Conde et al., Noble metal nanoparticles for biosensing applications. Sensors **12**(2), 1657–1687 (2012)

V.I. Dos Santos, C. Merlini et al., In vitro evaluation of bilayer membranes of PLGA/hydroxyapatite/β-tricalcium phosphate for guided bone regeneration. Mater. Sci. Eng.: C **112**, 110849 (2020)

A. Dzedzickis, V. Bucinskas et al., Modification of the AFM sensor by a precisely regulated air stream to increase imaging speed and accuracy in the contact mode. Sensors **18**(8), 2694 (2018)

C. Egbuna, J. Ifemeje, Oxidative stress and nutrition. Trop. J. Appl. Nat. Sci **2**(1), 110–116 (2017)

C. Egbuna, V.K. Parmar, et al., Toxicity of nanoparticles in biomedical application: nanotoxicology. J. Toxicol. (2021)

A. ElMekawy, H. Hegab et al., Bio-analytical applications of microbial fuel cell–based biosensors for onsite water quality monitoring. J. Appl. Microbiol. **124**(1), 302–313 (2018)

A.S. Emrani, N.M. Danesh et al., Colorimetric and fluorescence quenching aptasensors for detection of streptomycin in blood serum and milk based on double-stranded DNA and gold nanoparticles. Food Chem. **190**, 115–121 (2016)

M. Enea, E. Pereira et al., Gold nanoparticles induce oxidative stress and apoptosis in human kidney cells. Nanomaterials **10**(5), 995 (2020)

C. Fabbro, H. Ali-Boucetta et al., Targeting carbon nanotubes against cancer. Chem. Commun. **48**(33), 3911–3926 (2012)

Y. Fang, X. Zhu et al., Biodegradable core-shell electrospun nanofibers based on PLA and γ-PGA for wound healing. Eur. Polymer J. **116**, 30–37 (2019)

A.H. Faraji, P. Wipf, Nanoparticles in cellular drug delivery. Bioorg. Med. Chem. **17**(8), 2950–2962 (2009)

M. Faraji, Y. Yamini et al., Magnetic nanoparticles: synthesis, stabilization, functionalization, characterization, and applications. J. Iran. Chem. Soc. **7**(1), 1–37 (2010)

M. Fellah, N. Hezil et al., Preliminary investigation on the bio-tribocorrosion behavior of porous nanostructured β-type titanium based biomedical alloys. Mater. Lett. **257**, 126755 (2019)

O.S. Fenton, K.N. Olafson et al., Advances in biomaterials for drug delivery. Adv. Mater. **30**(29), 1705328 (2018)

D. Fernando, S. Sulthana et al., Cellular uptake and cytotoxicity of varying aspect ratios of gold nanorods in HeLa cells. ACS Appl. Bio Mater. **3**(3), 1374–1384 (2020)

R. Feynman, *There's plenty of room at the bottom* (CRC Press, Feynman and computation, 2018), pp. 63–76

A. Fraczek, E. Menaszek et al., Comparative in vivo biocompatibility study of single-and multi-wall carbon nanotubes. Acta Biomater. **4**(6), 1593–1602 (2008)

P. Ganguly, A. Breen et al., Toxicity of nanomaterials: exposure, pathways, assessment, and recent advances. ACS Biomater. Sci. Eng. **4**(7), 2237–2275 (2018)

L. Gao, Q. Yang et al., Recent advances in nanomaterial-enhanced enzyme-linked immunosorbent assays. Analyst **145**(12), 4069–4078 (2020)

T. García-Mendiola, I. Bravo et al., Carbon nanodots based biosensors for gene mutation detection. Sens. Actuators, B Chem. **256**, 226–233 (2018)

B. Gholamine, I. Karimi et al., Neurobehavioral toxicity of carbon nanotubes in mice: focus on brain-derived neurotrophic factor messenger RNA and protein. Toxicol. Ind. Health **33**(4), 340–350 (2017)

R. Gilchrist, R. Medal et al., Selective inductive heating of lymph nodes. Ann. Surg. **146**(4), 596 (1957)

E.M. Goncalves, F.J. Oliveira et al., Three-dimensional printed PCL-hydroxyapatite scaffolds filled with CNT s for bone cell growth stimulation. J. Biomed. Mater. Res. B Appl. Biomater. **104**(6), 1210–1219 (2016)

V.H. Grassian, P.T. O'Shaughnessy et al., Inhalation exposure study of titanium dioxide nanoparticles with a primary particle size of 2 to 5 nm. Environ. Health Perspect. **115**(3), 397–402 (2007)
J.-R. Gurr, A.S. Wang et al., Ultrafine titanium dioxide particles in the absence of photoactivation can induce oxidative damage to human bronchial epithelial cells. Toxicology **213**(1–2), 66–73 (2005)
S. Halivni, S. Shemesh et al., Inkjet printed fluorescent nanorod layers exhibit superior optical performance over quantum dots. Nanoscale **7**(45), 19193–19200 (2015)
B.A. Hammer, Y. Wu et al., Controlling cellular uptake and toxicity of polyphenylene dendrimers by chemical functionalization. ChemBioChem **18**(10), 960–964 (2017)
R.D. Handy, R. Owen et al., The ecotoxicology of nanoparticles and nanomaterials: current status, knowledge gaps, challenges, and future needs. Ecotoxicology **17**(5), 315–325 (2008)
J.B. Haun, T.J. Yoon et al., Magnetic nanoparticle biosensors. Wiley Interdisc. Rev.: Nanomed. Nanobiotechnol. **2**(3), 291–304 (2010)
L. He, S. Liao et al., Synergistic effects of electrospun PLLA fiber dimension and pattern on neonatal mouse cerebellum C17. 2 stem cells. Acta Biomater. **6**(8), 2960–2969 (2010)
J. Hengstler, Re: DNA damage induction and DNA repair inhibition prove co-exposures to cadmium, cobalt and lead as more dangerous than hitherto expected. Carcinogenesis **24**(11), 1855–1857 (2003)
P.H. Hoet, I. Brüske-Hohlfeld et al., Nanoparticles–known and unknown health risks. J. Nanobiotechnol. **2**(1), 1–15 (2004)
M. Horie, K. Nishio et al., Protein adsorption of ultrafine metal oxide and its influence on cytotoxicity toward cultured cells. Chem. Res. Toxicol. **22**(3), 543–553 (2009)
M. Horie, K. Nishio et al., Cellular responses by stable and uniform ultrafine titanium dioxide particles in culture-medium dispersions when secondary particle size was 100 nm or less. Toxicol. in Vitro **24**(6), 1629–1638 (2010)
M. Hosseini, H. Khabbaz et al., Aptamer-based colorimetric and chemiluminescence detection of aflatoxin B1 in foods samples. Acta Chim. Slov. **62**(3), 721–728 (2015)
S. Hossen, M.K. Hossain et al., Smart nanocarrier-based drug delivery systems for cancer therapy and toxicity studies: a review. J. Adv. Res. **15**, 1–18 (2019)
J. Hou, H. Liu et al., Mechanism of toxic effects of Nano-ZnO on cell cycle of zebrafish (Danio rerio). Chemosphere **229**, 206–213 (2019)
Z. Huaizhi, N. Yuantao, China's ancient gold drugs. Gold Bull. **34**(1), 24–29 (2001)
L. Huang, J. Hu et al., Nanomaterial applications for neurological diseases and central nervous system injury. Prog. Neurobiol. **157**, 29–48 (2017)
X. Huang, I.H. El-Sayed et al., Cancer cells assemble and align gold nanorods conjugated to antibodies to produce highly enhanced, sharp, and polarized surface Raman spectra: a potential cancer diagnostic marker. Nano Lett. **7**(6), 1591–1597 (2007)
X. Huang, L. Li et al., The shape effect of mesoporous silica nanoparticles on biodistribution, clearance, and biocompatibility in vivo. ACS Nano **5**(7), 5390–5399 (2011)
T. Huo, R.F. Barth et al., Preparation, biodistribution and neurotoxicity of liposomal cisplatin following convection enhanced delivery in normal and F98 glioma bearing rats. PLoS ONE **7**(11), e48752 (2012)
S. Hussain, K. Hess et al., In vitro toxicity of nanoparticles in BRL 3A rat liver cells. Toxicol. in Vitro **19**(7), 975–983 (2005)
S. Iijima, Helical microtubules of graphitic carbon. Nature **354**(6348), 56–58 (1991)
A. Ito, M. Shinkai et al., Medical application of functionalized magnetic nanoparticles. J. Biosci. Bioeng. **100**(1), 1–11 (2005)
J. Iturri, J.L. Toca-Herrera, Characterization of cell scaffolds by atomic force microscopy. Polymers **9**(8), 383 (2017)
H. Jahangirian, E.G. Lemraski et al., A review of drug delivery systems based on nanotechnology and green chemistry: green nanomedicine. Int. J. Nanomed. **12**, 2957 (2017)
H. Jawad, A.R. Lyon, et al. Myocardial tissue engineering. British Med. Bull. **87** (1) (2008)

Y. Jiang, H. Zhang et al., Modulation of apoptotic pathways of macrophages by surface-functionalized multi-walled carbon nanotubes. PLoS ONE **8**(6), e65756 (2013)
L. Jiao, L. Zhang et al., Au@ Pt nanodendrites enhanced multimodal enzyme-linked immunosorbent assay. Nanoscale **11**(18), 8798–8802 (2019)
M. Johannsen, U. Gneveckow et al., Thermotherapy of prostate cancer using magnetic nanoparticles: feasibility, imaging, and three-dimensional temperature distribution. Eur. Urol. **52**(6), 1653–1662 (2007)
H.J. Johnston, G. Hutchison et al., A review of the in vivo and in vitro toxicity of silver and gold particulates: particle attributes and biological mechanisms responsible for the observed toxicity. Crit. Rev. Toxicol. **40**(4), 328–346 (2010)
A. Jordan, Thermotherapy and nanomedicine, in *Hyperthermia in Cancer Treatment: A Primer* (Springer, 2006), pp. 60–63
A. Jordan, R. Scholz et al., Presentation of a new magnetic field therapy system for the treatment of human solid tumors with magnetic fluid hyperthermia. J. Magn. Magn. Mater. **225**(1–2), 118–126 (2001)
A. Jordan, R. Scholz et al., The effect of thermotherapy using magnetic nanoparticles on rat malignant glioma. J. Neurooncol. **78**(1), 7–14 (2006)
S.R. Joshi, A. Sharma et al., Low cost synthesis of reduced graphene oxide using biopolymer for influenza virus sensor. Mater. Sci. Eng.: C **108**, 110465 (2020)
D. Jović, V. Jaćević et al., The puzzling potential of carbon nanomaterials: general properties, application, and toxicity. Nanomaterials **10**(8), 1508 (2020)
L. Kaiser, The future of multihospital systems. Top. Health Care Financ. **18**(4), 32–45 (1992)
S. Kang, J.-E. Kim et al., Comparison of cellular toxicity between multi-walled carbon nanotubes and onion-like shell-shaped carbon nanoparticles. J. Nanopart. Res. **17**(9), 1–11 (2015)
M.N. Karim, J.E. Lee et al., Amperometric detection of catechol using tyrosinase modified electrodes enhanced by the layer-by-layer assembly of gold nanocubes and polyelectrolytes. Biosens. Bioelectron. **61**, 147–151 (2014)
N. Kobayashi, H. Izumi et al., Review of toxicity studies of carbon nanotubes. J. Occup. Health **17**, 0089 (2017)
F.-Y. Kong, J.-W. Zhang et al., Unique roles of gold nanoparticles in drug delivery, targeting and imaging applications. Molecules **22**(9), 1445 (2017)
P. Kovvuru, P.E. Mancilla et al., Oral ingestion of silver nanoparticles induces genomic instability and DNA damage in multiple tissues. Nanotoxicology **9**(2), 162–171 (2015)
C. Krishnaraj, S.L. Harper et al., In Vivo toxicological assessment of biologically synthesized silver nanoparticles in adult Zebrafish (Danio rerio). J. Hazard. Mater. **301**, 480–491 (2016)
A. Kroll, M.H. Pillukat et al., Current in vitro methods in nanoparticle risk assessment: limitations and challenges. Eur. J. Pharm. Biopharm. **72**(2), 370–377 (2009)
J. Kudr, Y. Haddad et al., Magnetic nanoparticles: from design and synthesis to real world applications. Nanomaterials **7**(9), 243 (2017)
M.K. Kumawat, M. Thakur et al., Graphene quantum dots for cell proliferation, nucleus imaging, and photoluminescent sensing applications. Sci. Rep. **7**(1), 1–16 (2017)
S. Kunjachan, A. Detappe et al., Nanoparticle mediated tumor vascular disruption: a novel strategy in radiation therapy. Nano Lett. **15**(11), 7488–7496 (2015)
S. Kurbanoglu, C. Erkmen et al., Frontiers in electrochemical enzyme based biosensors for food and drug analysis. TrAC Trends Anal. Chem. **124**, 115809 (2020)
S. Kurbanoglu, S.A. Ozkan et al., Nanomaterials-based enzyme electrochemical biosensors operating through inhibition for biosensing applications. Biosens. Bioelectron. **89**, 886–898 (2017)
P.P. LeDuc, R.R. Bellin, Nanoscale intracellular organization and functional architecture mediating cellular behavior. Ann. Biomed. Eng. **34**(1), 102–113 (2006)
H. Li, L. Chen et al., Ionic liquid-functionalized fluorescent carbon nanodots and their applications in electrocatalysis, biosensing, and cell imaging. Langmuir **30**(49), 15016–15021 (2014)

N. Li, X. Fan et al., Nanocomposite scaffold with enhanced stability by hydrogen bonds between collagen, polyvinyl pyrrolidone and titanium dioxide. Colloids Surf., B **140**, 287–296 (2016)

S. Li, S. Zhou et al., Exceptionally high payload of the IR780 iodide on folic acid-functionalized graphene quantum dots for targeted photothermal therapy. ACS Appl. Mater. Interfaces. **9**(27), 22332–22341 (2017)

W. Li, X. Zhang et al., Shape design of high drug payload nanoparticles for more effective cancer therapy. Chem. Commun. **49**(93), 10989–10991 (2013)

Y.-C.E. Li, I. Lee, The current trends of biosensors in tissue engineering. Biosensors **10**(8), 88 (2020)

Z. Li, T. Hulderman et al., Cardiovascular effects of pulmonary exposure to single-wall carbon nanotubes. Environ. Health Perspect. **115**(3), 377–382 (2007)

H. Liao, C.L. Nehl, et al., Biomedical applications of plasmon resonant metal nanoparticles (2006)

Q. Lin, X. Hong et al., Biosynthesis of size-controlled gold nanoparticles using M. lucida leaf extract and their penetration studies on human skin for plastic surgery applications. J. Photochem. Photobio. b: Bio. **199**, 111591 (2019)

M. Lippmann, Effects of fiber characteristics on lung deposition, retention, and disease. Environ. Health Perspect. **88**, 311–317 (1990)

G. Liu, M. Fu et al., Tissue-engineered PLLA/gelatine nanofibrous scaffold promoting the phenotypic expression of epithelial and smooth muscle cells for urethral reconstruction. Mater. Sci. Eng.: C **111**, 110810 (2020)

B.E. Logan, R. Rossi et al., Electroactive microorganisms in bioelectrochemical systems. Nat. Rev. Microbiol. **17**(5), 307–319 (2019)

C. Lopez-Chaves, J. Soto-Alvaredo, et al., Gold nanoparticles: distribution, bioaccumulation and toxicity. In vitro and in vivo studies. Nanomed.: Nanotechnol. Bio. Med. **14** (1), 1–12 (2018)

G. Maduraiveeran, W. Jin, Functional nanomaterial-derived electrochemical sensor and biosensor platforms for biomedical applications, in *Handbook of Nanomaterials in Analytical Chemistry* (Elsevier, 2020), pp. 297–327

M. Mahmoudi, M.A. Shokrgozar et al., Irreversible changes in protein conformation due to interaction with superparamagnetic iron oxide nanoparticles. Nanoscale **3**(3), 1127–1138 (2011)

K. Maier-Hauff, F. Ulrich et al., Efficacy and safety of intratumoral thermotherapy using magnetic iron-oxide nanoparticles combined with external beam radiotherapy on patients with recurrent glioblastoma multiforme. J. Neurooncol. **103**(2), 317–324 (2011)

D. Maiti, X. Tong et al., Carbon-based nanomaterials for biomedical applications: a recent study. Front. Pharmacol. **9**, 1401 (2019)

A. Manke, L. Wang, et al., Mechanisms of nanoparticle-induced oxidative stress and toxicity. BioMed. Res. Int. (2013)

E. Martínez-Periñán, I. Bravo et al., Carbon nanodots as electrocatalysts towards the oxygen reduction reaction. Electroanalysis **30**(3), 436–444 (2018)

D. Mateo, P. Morales et al., Oxidative stress contributes to gold nanoparticle-induced cytotoxicity in human tumor cells. Toxicol. Mech. Methods **24**(3), 161–172 (2014)

V. Mazzaracchio, M.R. Tomei et al., Inside the different types of carbon black as nanomodifiers for screen-printed electrodes. Electrochim. Acta **317**, 673–683 (2019)

K. McNamara, S.A. Tofail, Nanoparticles in biomedical applications. Adv. Phys.: X **2**(1), 54–88 (2017)

A. Mehdizadeh, S. Pandesh et al., The effects of folate-conjugated gold nanorods in combination with plasmonic photothermal therapy on mouth epidermal carcinoma cells. Lasers Med. Sci. **29**(3), 939–948 (2014)

S.K. Mohanty, M.K. Swamy et al., Leptadenia reticulata (Retz.) Wight & Arn. (Jivanti): botanical, agronomical, phytochemical, pharmacological, and biotechnological aspects. Molecules **22**(6), 1019 (2017)

S. Mornet, S. Vasseur et al., Magnetic nanoparticle design for medical applications. Prog. Solid State Chem. **34**(2–4), 237–247 (2006)

E.A. Morsy, A.M. Hussien, et al., Cytotoxicity and genotoxicity of copper oxide nanoparticles in chickens. Biol. Trace Elem. Res. 1–15 (2021)

C.E. Nichols, D.L. Shepherd et al., Reactive oxygen species damage drives cardiac and mitochondrial dysfunction following acute nano-titanium dioxide inhalation exposure. Nanotoxicology **12**(1), 32–48 (2018)

S.K. Nune, P. Gunda et al., Nanoparticles for biomedical imaging. Expert Opin. Drug Deliv. **6**(11), 1175–1194 (2009)

G. Oberdörster, Safety assessment for nanotechnology and nanomedicine: concepts of nanotoxicology. J. Intern. Med. **267**(1), 89–105 (2010)

G. Oberdörster, E. Oberdörster et al., Nanotoxicology: an emerging discipline evolving from studies of ultrafine particles. Environ. Health Perspect. **113**(7), 823–839 (2005)

G. Oberdörster, V. Stone et al., Toxicology of nanoparticles: a historical perspective. Nanotoxicology **1**(1), 2–25 (2007)

A.L. Oldenburg, M.N. Hansen et al., Imaging gold nanorods in excised human breast carcinoma by spectroscopic optical coherence tomography. J. Mater. Chem. **19**(35), 6407–6411 (2009)

T. Panczyk, P. Wolski et al., Coadsorption of doxorubicin and selected dyes on carbon nanotubes. Theoretical investigation of potential application as a pH-controlled drug delivery system. Langmuir **32**(19), 4719–4728 (2016)

J. Pandiarajan, S. Balaji, Genotoxic effect of silver nanoparticles in silk worm Bombyx mori. Austin J. Biotechnol. Bioeng **5**, 1096 (2018)

H.J. Parab, C. Jung et al., A gold nanorod-based optical DNA biosensor for the diagnosis of pathogens. Biosens. Bioelectron. **26**(2), 667–673 (2010)

D. Patel, B. Rajeswari et al., Structural, luminescence and EPR studies on SrSnO 3 nanorods doped with europium ions. Dalton Trans. **41**(39), 12023–12030 (2012)

D.K. Patel, B. Vishwanadh et al., Difference in the nature of Eu 3+ environment in Eu 3+-doped BaTiO 3 and BaSnO 3. J. Am. Ceram. Soc. **96**(12), 3857–3861 (2013)

N.H. Patil, P.V. Devarajan, Insulin-loaded alginic acid nanoparticles for sublingual delivery. Drug Deliv. **23**(2), 429–436 (2016)

V.P. Pattani, J.W. Tunnell, Nanoparticle-mediated photothermal therapy: a comparative study of heating for different particle types. Lasers Surg. Med. **44**(8), 675–684 (2012)

F. Pederzoli, G. Tosi et al., Protein corona and nanoparticles: how can we investigate on? Wiley Interdisc. Rev.: Nanomed. Nanobiotechnol. **9**(6), e1467 (2017)

J. Peña-Bahamonde, H.N. Nguyen et al., Recent advances in graphene-based biosensor technology with applications in life sciences. J. Nanobiotechnol. **16**(1), 1–17 (2018)

Z. Peng, X. Liu et al., Advances in the application, toxicity and degradation of carbon nanomaterials in environment: a review. Environ. Int. **134**, 105298 (2020)

V. Perumal, U. Hashim, Advances in biosensors: principle, architecture and applications. J. Appl. Biomed. **12**(1), 1–15 (2014)

J.M. Pettibone, A. Adamcakova-Dodd et al., Inflammatory response of mice following inhalation exposure to iron and copper nanoparticles. Nanotoxicology **2**(4), 189–204 (2008)

V. Pokropivny, V. Skorokhod, Classification of nanostructures by dimensionality and concept of surface forms engineering in nanomaterial science. Mater. Sci. Eng., C **27**(5–8), 990–993 (2007)

A. Portero, C. Remunan-Lopez et al., Reacetylated chitosan microspheres for controlled delivery of anti-microbial agents to the gastric mucosa. J. Microencapsul. **19**(6), 797–809 (2002)

D. Pozzi, V. Colapicchioni et al., Effect of polyethyleneglycol (PEG) chain length on the bio–nano-interactions between PEGylated lipid nanoparticles and biological fluids: from nanostructure to uptake in cancer cells. Nanoscale **6**(5), 2782–2792 (2014)

A. Prévoteau, K. Rabaey, Electroactive biofilms for sensing: reflections and perspectives. ACS Sens. **2**(8), 1072–1085 (2017)

K. Prusty, S.K. Swain, Nano silver decorated polyacrylamide/dextran nanohydrogels hybrid composites for drug delivery applications. Mater. Sci. Eng., C **85**, 130–141 (2018)

M. Rai, A. Yadav et al., Silver nanoparticles as a new generation of antimicrobials. Biotechnol. Adv. **27**(1), 76–83 (2009)

S. Rawal, M.M. Patel, Threatening cancer with nanoparticle aided combination oncotherapy. J. Control. Release **301**, 76–109 (2019)

M. Razzaghi, M. Kasiri-Asgarani et al., Microstructure, mechanical properties, and in-vitro biocompatibility of nano-NiTi reinforced Mg–3Zn-0.5 Ag alloy: prepared by mechanical alloying for implant applications. Compos. Part b: Eng. **190**, 107947 (2020)

E. Reimhult, Nanoparticle-triggered release from lipid membrane vesicles. New Biotechnol. **32**(6), 665–672 (2015)

S.A. Rizvi, A.M. Saleh, Applications of nanoparticle systems in drug delivery technology. Saudi Pharm. J. **26**(1), 64–70 (2018)

T.G. Rocha, H.F.L. Pedro et al., Lipase cocktail for optimized biodiesel production of free fatty acids from residual chicken oil. Catal. Lett. **151**(4), 1155–1166 (2021)

N.J. Ronkainen, H.B. Halsall et al., Electrochemical biosensors. Chem. Soc. Rev. **39**(5), 1747–1763 (2010)

S.V. Roussakow, A randomized clinical trial of radiation therapy versus thermoradiotherapy in stage IIIB cervical carcinoma of Yoko Harima et al. (2001): multiple biases and no advantage of hyperthermia. Int. J. Hyperth. **34**, 1400 (2018)

C. Rümenapp, B. Gleich et al., Magnetic nanoparticles in magnetic resonance imaging and diagnostics. Pharm. Res. **29**(5), 1165–1179 (2012)

J.H. Ryu, H.Y. Yoon et al., Tumor-targeting glycol chitosan nanoparticles for cancer heterogeneity. Adv. Mater. **32**(51), 2002197 (2020)

T.A. Saleh, Nanomaterials: classification, properties, and environmental toxicities. Environ. Technol. Innov. **20**, 101067 (2020)

D.A. Saud Alarifi, Y. Al Omar Suliman et al., Oxidative stress contributes to cobalt oxide nanoparticles-induced cytotoxicity and DNA damage in human hepatocarcinoma cells. Int. J. Nanomed. **8**, 189 (2013)

J.A. Schwartz, R.E. Price, et al., Selective nanoparticle-directed photothermal ablation of the canine prostate, in *Energy-based Treatment of Tissue and Assessment VI, International Society for Optics and Photonics* (2011)

M.R.M. Shafiee, M. Kargar, Preparation of aryl sulfonamides using CuO nanoparticles prepared in extractive Rosmarinus Officinalis leaves media. Biointerface Res. Appl. Chem. **6** (3) (2016)

D. Shi, G. Mi, et al., The synthesis, application, and related neurotoxicity of carbon nanotubes, in *Neurotoxicity of Nanomaterials and Nanomedicine* (Elsevier, 2017), pp. 259–284

M. Shinkai, M. Yanase et al., Intracellular hyperthermia for cancer using magnetite cationic liposomes: in vitro study. Jpn. J. Cancer Res. **87**(11), 1179–1183 (1996)

F. Siepmann, S. Herrmann et al., A novel mathematical model quantifying drug release from lipid implants. J. Control. Release **128**(3), 233–240 (2008)

H. Sies, Oxidative stress: from basic research to clinical application. Am. J. Med. **91**(3), S31–S38 (1991)

S. Sim, N.K. Wong, Nanotechnology and its use in imaging and drug delivery. Biomed. Rep. **14**(5), 1–9 (2021)

A.M.S. Simão, M. Bolean et al., Liposomal systems as carriers for bioactive compounds. Biophys. Rev. **7**(4), 391–397 (2015)

P.P. Simeonova, A. Erdely, Engineered nanoparticle respiratory exposure and potential risks for cardiovascular toxicity: predictive tests and biomarkers. Inhalation Toxicol. **21**(sup, 1), 68–73 (2009)

N. Singh, M.A. Ali et al., Microporous nanocomposite enabled microfluidic biochip for cardiac biomarker detection. ACS Appl. Mater. Interfaces. **9**(39), 33576–33588 (2017)

W. Song, J. Zhang et al., Role of the dissolved zinc ion and reactive oxygen species in cytotoxicity of ZnO nanoparticles. Toxicol. Lett. **199**(3), 389–397 (2010)

S. Sruthi, L. Maurizi et al., Cellular interactions of functionalized superparamagnetic iron oxide nanoparticles on oligodendrocytes without detrimental side effects: Cell death induction, oxidative stress and inflammation. Colloids Surf., B **170**, 454–462 (2018)

K.P. Steckiewicz, E. Barcinska et al., Impact of gold nanoparticles shape on their cytotoxicity against human osteoblast and osteosarcoma in in vitro model. Evaluation of the safety of use and anti-cancer potential. J. Mater. Sci. Mater. Med. **30**(2), 22 (2019)
B. Sun, M. Zhang et al., Applications of cellulose-based materials in sustained drug delivery systems. Curr. Med. Chem. **26**(14), 2485–2501 (2019)
C. Sun, N. Yin et al., Silver nanoparticles induced neurotoxicity through oxidative stress in rat cerebral astrocytes is distinct from the effects of silver ions. Neurotoxicology **52**, 210–221 (2016)
J. Tang, X. Lu et al., Mechanisms of silver nanoparticles-induced cytotoxicity and apoptosis in rat tracheal epithelial cells. J. Toxicol. Sci. **44**(3), 155–165 (2019)
Z. Tang, C. He et al., Polymeric nanostructured materials for biomedical applications. Prog. Polym. Sci. **60**, 86–128 (2016)
D.R. Thévenot, K. Toth et al., Electrochemical biosensors: recommended definitions and classification. Biosens. Bioelectron. **16**(1–2), 121–131 (2001)
C. Tian, L. Zhu et al., Poly (acrylic acid) bridged gadolinium metal–organic framework–gold nanoparticle composites as contrast agents for computed tomography and magnetic resonance bimodal imaging. ACS Appl. Mater. Interfaces. **7**(32), 17765–17775 (2015)
C. Velez, D.K. Patel et al., Hierarchical integration of thin-film NiTi actuators using additive manufacturing for microrobotics. J. Microelectromech. Syst. **29**(5), 867–873 (2020)
S. Vieira, S. Vial et al., Nanoparticles for bone tissue engineering. Biotechnol. Prog. **33**(3), 590–611 (2017)
E. Villar-Alvarez, A. Cambón et al., Gold nanorod-based nanohybrids for combinatorial therapeutics. ACS Omega **3**(10), 12633–12647 (2018)
D. Wang, T. Imae et al., Fluorescence emission from PAMAM and PPI dendrimers. J. Colloid Interface Sci. **306**(2), 222–227 (2007)
H. Wang, T.B. Huff et al., In vitro and in vivo two-photon luminescence imaging of single gold nanorods. Proc. Natl. Acad. Sci. **102**(44), 15752–15756 (2005)
J. Wang, S. Cao et al., Theoretical investigations of optical origins of fluorescent graphene quantum dots. Sci. Rep. **6**(1), 1–5 (2016)
J. Wang, A. Shi et al., An ultrasensitive supersandwich electrochemical DNA biosensor based on gold nanoparticles decorated reduced graphene oxide. Anal. Biochem. **469**, 71–75 (2015)
L. Wang, S. Zhang et al., In-vitro antibacterial and anti-encrustation performance of silver-polytetrafluoroethylene nanocomposite coated urinary catheters. J. Hosp. Infect. **103**(1), 55–63 (2019)
R. Watkins, L. Wu et al., Natural product-based nanomedicine: recent advances and issues. Int. J. Nanomed. **10**, 6055 (2015)
C.-J. Wen, L.-W. Zhang et al., Theranostic liposomes loaded with quantum dots and apomorphine for brain targeting and bioimaging. Int. J. Nanomed. **7**, 1599 (2012)
K.C.-W. Wu, Y. Yamauchi et al., Biocompatible, surface functionalized mesoporous titania nanoparticles for intracellular imaging and anticancer drug delivery. Chem. Commun. **47**(18), 5232–5234 (2011)
T. Wu, T. Zhang et al., Research advances on potential neurotoxicity of quantum dots. J. Appl. Toxicol. **36**(3), 345–351 (2016)
P. Wust, B. Hildebrandt et al., Hyperthermia in combined treatment of cancer. Lancet Oncol. **3**(8), 487–497 (2002)
R. Xie, Z. Wang et al., Graphene quantum dots as smart probes for biosensing. Anal. Methods **8**(20), 4001–4016 (2016)
Y. Xue, T. Zhang et al., Cytotoxicity and apoptosis induced by silver nanoparticles in human liver HepG2 cells in different dispersion media. J. Appl. Toxicol. **36**(3), 352–360 (2016)
K. Yang, L. Feng et al., Nano-graphene in biomedicine: theranostic applications. Chem. Soc. Rev. **42**(2), 530–547 (2013)
J. Yao, P. Li et al., Biochemistry and biomedicine of quantum dots: from biodetection to bioimaging, drug discovery, diagnostics, and therapy. Acta Biomater. **74**, 36–55 (2018)

T. Yin, P. Wang et al., Ultrasound-sensitive siRNA-loaded nanobubbles formed by hetero-assembly of polymeric micelles and liposomes and their therapeutic effect in gliomas. Biomaterials **34**(18), 4532–4543 (2013)

Z. Yin, D. Li et al., Regeneration of elastic cartilage with accurate human-ear shape based on PCL strengthened biodegradable scaffold and expanded microtia chondrocytes. Appl. Mater. Today **20**, 100724 (2020)

R. Yuan, T. Rao et al., Quantum dot-based fluorescent probes for targeted imaging of the EJ human bladder urothelial cancer cell line. Exp. Ther. Med. **16**(6), 4779–4783 (2018)

E.N. Zare, P. Makvandi et al., Progress in conductive polyaniline-based nanocomposites for biomedical applications: a review. J. Med. Chem. **63**(1), 1–22 (2019)

X.Q. Zhang, R. Lam et al., Multimodal nanodiamond drug delivery carriers for selective targeting, imaging, and enhanced chemotherapeutic efficacy. Adv. Mater. **23**(41), 4770–4775 (2011)

F. Zhou, X. Da et al., Cancer photothermal therapy in the near-infrared region by using single-walled carbon nanotubes. J. Biomed. Opt. **14**(2), 021009 (2009)

Nanomaterials and Purification Techniques for Water Purification and Wastewater Treatment

Twinkle Twinkle, Krati Saini, Ravi K. Shukla, Achintya N. Bezbaruah, Rajeev Gupta, Kamal K. Kar, K. K. Raina, and Pankaj Chamoli

Abstract Access to safe and clean water has become a more challenging task worldwide as water resources are limited, and the population that relies on these limited supplies is expected to grow. The presence of pollutants in water affects human health and hygiene and decreases food safety. The supply of clean water is required for all the phases of food production, including processing, transportation, and consumption. Environmentally viable nanomaterials are being used to purify wastewater because of their distinctive characteristics, like high effectiveness and selectivity, larger surface area, cost-effective, recyclable, and high thermal and mechanical stability. This chapter provides an overall review of nanomaterials and their types and techniques used to eliminate organic and inorganic contaminants from wastewater.

Keywords Nanomaterials · Carbon nanomaterials · Photocatalyst · Absorbent · Water purification

---

T. Twinkle · K. Saini · P. Chamoli (✉)
School of Basic and Applied Sciences, Department of Physics, Shri Guru Ram Rai University, Dehradun, Uttarakhand 248001, India
e-mail: pankajchamoli@sgrru.ac.in; pchamoli83@gmail.com

R. K. Shukla
Advanced Functional Smart Materials Laboratory, School of Physical Sciences, Department of Physics, DIT University, Dehradun, Uttarakhand 248009, India

A. N. Bezbaruah
Nanoenvirology Research Group, Civil, Construction and Environmental Engineering, North Dakota State University, Fargo, ND 58105, USA

R. Gupta
Department of Physics, School of Engineering Studies, University of Petroleum and Energy Studies, Dehradun, Uttarakhand 248007, India

K. K. Kar
Advanced Nanoengineering Materials Laboratory, Materials Science Programme, Indian Institute of Technology Kanpur, Kanpur 208016, India

K. K. Raina
MS Ramaiah University of Applied Sciences, Bangalore Urban, Karnataka 560054, India

J. K. Katiyar et al. (eds.), *Nanomaterials for Advanced Technologies*,
https://doi.org/10.1007/978-981-19-1384-6_6

1 Introduction

Water is the most critical component for every living creature present in the world. Clean and drinkable water is a primary need for all living things and mankind on this earth. Fresh and clean water supply is a major challenge faced by human society and is expected to increase in the coming years. Availability of drinking water is far from the demand due to the gradual increase in population and various contaminates (Kunduru et al. 2017). A constant supply of safe and reliable water is required for proper sanitation for optimal health. The individual water requirements range from 7.5 L per day for drinking purposes and approximately 20–50 L per day for other daily needs (Khan and Malik 2019).

On the other hand, both developed and developing countries have reported waterborne diseases and believe that 50% of people in underdeveloped countries do not have access to clean and safe drinking water. Therefore, in developing countries, the uncontrolled discharge of contaminants that inadequacy of water purification system is the main problem (Figoli et al. 2017a; Borji et al. 2020). The causes of water pollution include improper sewage disposal, mining, industrial waste, oil spills, chemical fertilizers, pesticides, radioactive water discharge, etc. (Fig. 1a) (Kaur et al. 2020). The contaminants of wastewater affect our human health and cause several health problems like diarrhea, jaundice, impaired nerve function, skin infections, brain damage, dysfunction of the liver, etc. (Fig. 1b) (Cínti et al. 2019).

The prospect of the increasing freshwater source is limited because of the competing demands of the world's growing population. Industrial development improved the lifestyle of humans and damaged our natural resources and aquatic systems. Industries such as leather, food, pharmaceuticals, and packaging generate an enormous daily amount of wastewater which contains heavy metals, organic dyes, and other harmful chemicals. These gallons of wastewater are dumped into our

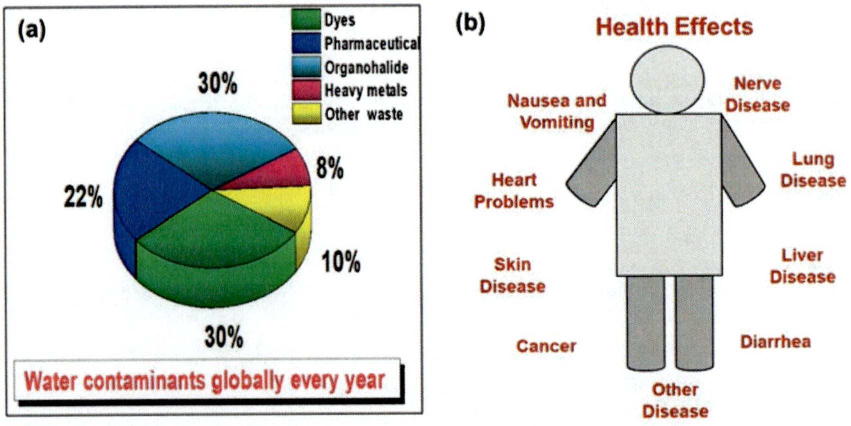

Fig. 1 **a** Global release of various pollutants into the water every year [Redrawn from Kaur et al. (2020)], **b** Possible effects of water contaminants on human health (Cínti et al. 2019)

aquatic ecosystem without any meaningful treatment. Thus, the effective removal of such contaminants is needed. Nanotechnology has shown the potential to solve water quality problems effectively due to its environment-friendly and non-toxicity properties for water purification. In particular, the vast range of nanostructures (NSs) such as metal oxide (ZnO, TiO_2, ZnS, Fe_3O_4, etc.) NSs [NPs (NPs), nanorods (NRs), etc.] and their nanocomposites (NCs), Carbon nanotubes (CNTs) or Carbon nanotube (CNT), graphene oxide (GO), reduced graphene oxide (rGO) graphene nanosheets (GNs) helps to develop a more effective treatment in advanced water purification systems (Amin et al. 2014). Unique advantages of nanomaterials like high surface-to-volume ratio, small size, well-organized structure, and ability of filtration make them a potential alternative for water treatment. Although, there are still some major challenges like high production cost, specific selectivity and availability, sustainability, and recyclability. Worldwide, various scientific groups are trying to develop highly effective and environmentally viable nanomaterials at a cost-effective for ecological purification of wastewater which contains hazardous heavy metals ions, dyes, detergents, and chemical waste from industries (Nasrollahzadeh et al. 2021).

In this chapter, we have discussed nanotechnology-enabled technologies that utilize different nanomaterials like silver NPs, metal-based NPs, carbon-based nanomaterials, etc. Their properties and reviewed other technologies for water purification adsorption disinfection, photocatalytic action, membranes, etc. Properties of various nano materials are reviewed in this chapter and varius technologies like water purification adsorption disinfection, photocatalytic action, membranes, etc. are discussed.

1.1 Nanomaterials for Water Purification

1.1.1 Zero-valent Metal NPs

Silver (Ag) is a transition metal (soft and shiny) whose atomic number is 47 (Fig. 2a) and exhibits greater electrical conductivity (~6.3×10^7 m/Ω) and thermal conductivity (~429 W/m K). Ag NPs are the most extensively utilized material because of their low toxicity and can be easily extracted from their salts, such as silver nitrate and silver chloride. Ag NPs have been used for various applications such as antibacterial, thin films, and water purification (Chamoli et al. 2017; Maninder and Baojun 2019). Ag NPs have strong microbial inactivation in water and show excellent antibacterial effects against various micro-organisms such as viruses, bacteria, and fungi. Therefore, it is commonly used for water disinfection due to its antibacterial properties (Lu et al. 2016). In particular, Ag NPs have been synthesized and utilized for the photodegradation of various organic dyes. For example, Pandey et al. synthesized Ag NPs from κ-Carrageenan gum and investigated their photocatalytic activity in the presence of UV with RhB and MB as the target pollutants. They were able to remove up to 100% in a short period of time (Pandey et al. 2020). Similarly, Rajkumar et al. have synthesized Ag NPs by utilizing cell-free extract of Chlorella Vulgaris. The

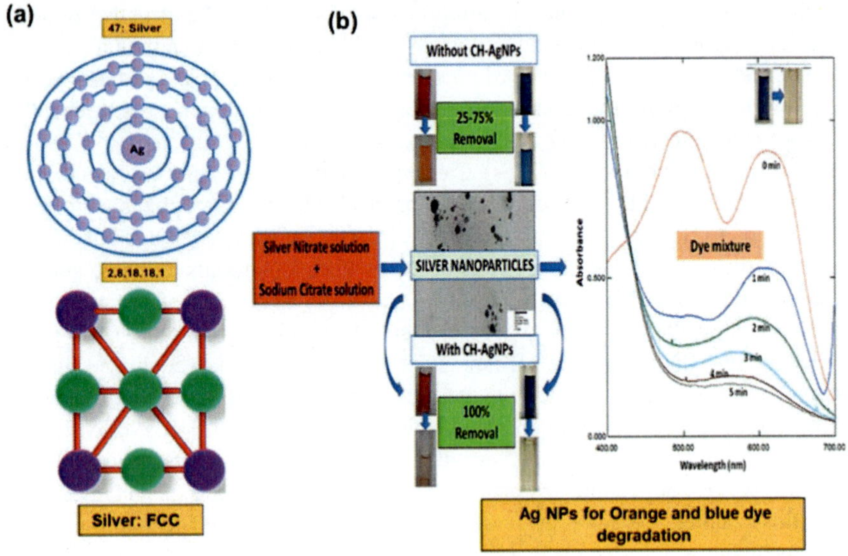

Fig. 2 **a** Crystal structures of Ag and **b** Ag NPs for orange and blue dye degradation. Reproduced with permission from Gola et al. (2021)

synthesized Ag NPs showed 96.51% of photocatalytic decolorization activity using methylene blue dye (100 ppm) within 3 h incubation time (Rajkumar et al. 2021). However, Gola et al. have synthesized Ag NPs and achieved degradation up to 100% from blue dye and orange dye ~97.4% degradation, respectively. Further, dye mixture studies (orange + blue dye) have been examined and found 100% degradation in just 5 min (Fig. 2b) (Gola et al. 2021). Jain et al. have prepared aqueous Ag NPs (Pa–Ag NPs) using leaf extract of C. papaya and studied dye degradation ability for blue CP and yellow 3RS with degradation ability 90 and 83%, respectively (Jain et al. 2020).

1.1.2 Metal Oxide NSs and NCs

TiO_2 NSs

Titanium dioxide (TiO_2) is a wide bandgap semiconductor with three different crystallographic forms (polymorphs, Fig. 3a), viz., anatase (with a bandgap of 3.1 eV), rutile (3.02 eV), and brookite (2.96 eV) (Haggerty et al. 2017). TiO_2-anatase is an extensively used photocatalyst due to its high photocatalytic activity, low price, and good biological and chemical stability. Under the presence of light (UV and visible), TiO_2 NSs act as an excellent photocatalyst and successfully degrade various organic contaminants. Primarily, TiO_2 produces many reactive oxygen species that can entirely deteriorate pollutants in a short reaction time under ultraviolet irradiation (Ali et al. 2018a).

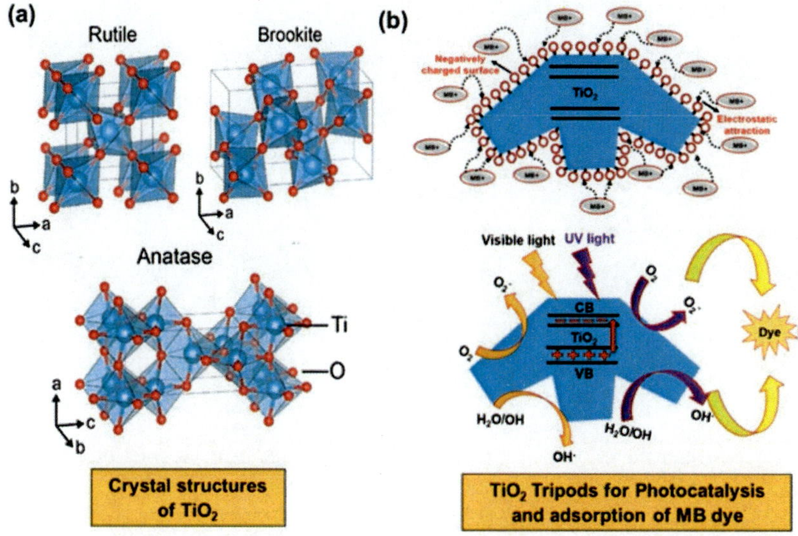

Fig. 3 **a** Crystal structures of TiO_2: rutile; brookite, and anatase; Reproduced with permission from Haggerty et al. (2017) and **b** TiO_2 tripods for photocatalysis and adsorption; Reproduced with permission from Chamoli et al. (2021a)

For example, Gautam et al. have prepared TiO_2 (both anatase and rutile) NPs and successfully degraded MB dye in the presence of UV (8 W) light irradiation, and the degradation efficiency (DE, ~88%) is obtained by using anatase NPs in 150 min (Gautama et al. 2016). Tayeb et al. have synthesized TiO2 NPs and effectively degraded MB dye under UV (15 W) light irradiation with degradation efficiency (DE) of ~ 98% in 90 min (Tayeb and Hussein 2015). Sathiyan et al. have prepared TiO_2 NPs and effectively degraded MB dye up to 88% in 180 min in the presence of UV–visible light (Sathiyana et al. 2020). Chamoli et al. used Mangifera indica leaf extract to make TiO_2 tripods (TiTPs) via a rapid microwave (180 s, 100 W) green method. TiTPs have shown excellent photocatalytic ability against MB, achieving dye degradation of ~75% (under visible light in 75 min) and 96% (under UV light in 9 min). Moreover, TiTPs have exhibited good adsorbent capabilities, with a maximum adsorption capacity ~17.54 mg/g based on the Langmuir model owing to their porous nature (Fig. 3b) (Chamoli et al. 2021a).

ZnO NPs

ZnO is a compound semiconductor material of group-II-VI. The majority of the materials in this group-II-VI are cubic zinc-blende or hexagonal wurtzite structures (four cations surround each anion at the corners of a tetrahedron). ZnO is a 3.37 eV broad bandgap semiconductor with the structure of wurtzite (B4), zinc-blende (B3), and rock salt (B1) (Fig. 4a), and its ionicity is intermediate between covalent and

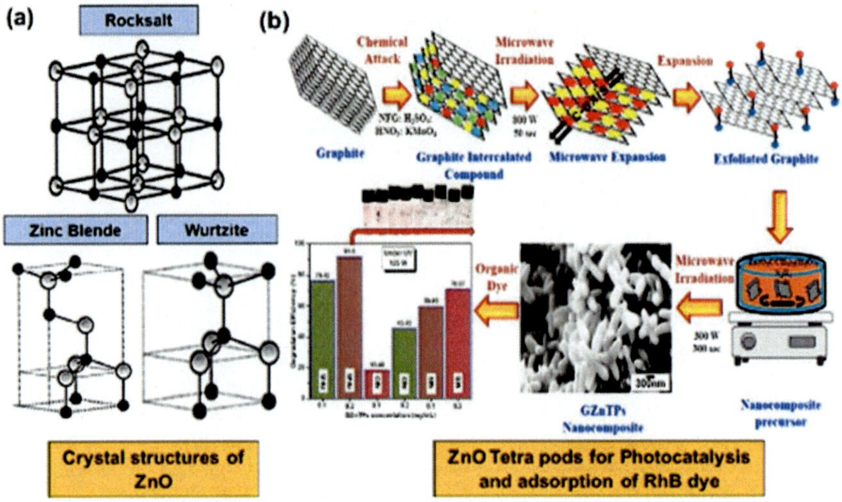

Fig. 4 **a** Stick and ball representation of ZnO crystal structures: cubic rock salt (B1), cubic zinc-blende (B3), and hexagonal wurtzite (B4). The shaded gray and black spheres denote Zn and O atoms, respectively. Reproduced with permission from Özgür et al. (2005). **b** GZnTPs for photocatalysis and adsorption of RhB dye, Reproduced with permission from Chamoli et al. (2021b)

ionic semiconductors (Özgür et al. 2005). ZnO is extensively employed in materials because of its ease of production and low toxicity. It is useful in wastewater remediation due to its distinctive features, such as direct and wide bandgap in the near-ultraviolet spectral region, high oxidation ability, and enhanced photocatalytic performance.

For example, Fan et al. have synthesized zinc oxide-reduced graphene oxide (ZnO/rGO) NCs for photocatalytic degradation of MB, MO, and RhB in the presence of ultraviolet (UV) light irradiation (150 W) and showed degradation efficiency (DE) of ~99% in 30 min (Fan et al. 2015). Ravi et al. have produced ZnO/rGO NCs for photocatalytic degradation of Congo red (CR) and eosin yellow (EY) in the presence of UV light and found 98% removal of the dyes in 90 min (Ravi et al. 2018). Jabeen et al. have attained 68% removal of methylene blue (MB) dye in 120 min in the presence of UV light (500 W) by employing ZnO/rGO NCs (Jabeen et al. 2017). Furthermore, for the removal of RhB, ternary ZnO/CuO/rGO NCs have also been produced, with 99% degradation efficiency (DE) obtained in 20 min in the presence of visible light (150 W) (Kumaresan et al. 2020). Chamoli et al. have synthesized grapheme—ZnO tetrapods (GZnTPs) and investigated their potential to photodegrade RhB, MO, and MB dyes in the presence of UV and visible (both 125 W) light irradiation. Upon UV light irradiation, GZnTPs behave as an outstanding photocatalyst for RhB, with such a higher degradation efficiency of 91.6% (Chamoli et al. 2021b). Compared to numerous semiconductor metallic oxides, ZnO NPs can adsorb a much broader solar spectral range and more light quanta. Like TiO_2 NPs, the light absorption of ZnO NPs is limited to the UV range due to its large band energy. In addition, the utility of ZnO particles is hindered by photo-corrosion, leading to the fact that the photo-generated charges recombine rapidly, resulting in lower photocatalytic efficiency.

Zinc Sulfide (ZnS) NSs

ZnS (II-VI group semiconductor) is an important photocatalyst studied at the nanoscale because of its outstanding physical properties and peculiar photocatalytic properties. Cubic (sphalerite) and hexagonal (wurtzite) are two main crystalline forms of ZnS with coordination geometry at Zn and S are of tetrahedral having 3.72 and 3.77 eV bandgaps for cubic and hexagonal ZnS, respectively, (Lee and Wu 2017). This large bandgap of ZnS enables candidacy as an important photocatalyst for various dye degradation and wastewater remediation. For example, Maji et al. have prepared ZnS NCs (rod and sphere) using ethylenediamine and hexadecylamine, which show effective photocatalytic activity against rose bengal dye (RB) under light irradiation. The DE has been achieved ~93% at 225 min (Maji et al. 2011). Zhang et al. have successfully prepared ZnS microcrystals (polyhedron, fan-shaped sheet, hexagonal rectangle, and missing angle rectangle) using a simple hydrothermal method against MB. The degradation efficiency (DE) has been achieved ~91% at 60 min (Zhang 2014). However, metal (Pb, Cu, Ni)-doped ZnS photocatalyst obtained by various methods, and also conjugated ZnS complexes obtained through polyreaction have been shown to have the potential to defluorinate hexafluorobenzene by visible light (Lee and Wu 2017). Thio-glycerol and uncapped ZnS NPs have been produced to use a substantial component of solar energy for dye degradation. Although ultraviolet irradiation is excellent at degrading dyes, naturally occurring solar radiation is also beneficial in dye degradation. As a result, it could be an effective approach for the safe disposal of textile waste into waterways (Sharma et al. 2012), while Lee et al. have been summarized ZnS-assisted photocatalytic degradation of pollutants and water splitting under various conditions (Lee and Wu 2017).

Fe_3O_4 NSs

In recent years, people have been more interested in using iron oxide NPs due to their simplicity and easy accessibility to degrade organic dyes and remove heavy metals from wastewater. Iron oxide NPs have different forms (Fig. 5) and are commonly used as nanoadsorbents such as magnetic magnetite (Fe_3O_4), magnetic maghemite (γ Fe_2O_4), and non-magnetic hematite (α Fe_2O_4). Separating and recovering them from wastewater during photocatalytic degradation is a challenging task. Thus, Fe_3O_4 and Fe_2O_4 can be easily be separated and recovered by using an external magnetic field from the wastewater system (Xu et al. 2012). Therefore, iron oxides are widely used in the purification of water.

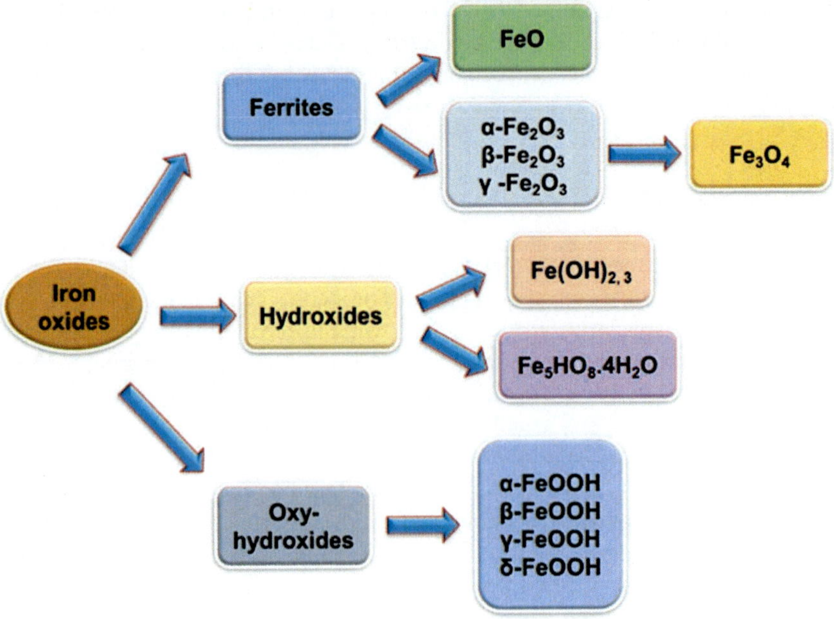

Fig. 5 Different forms of iron-based NPs: oxides, hydroxides, and oxyhydroxides (Aragaw et al. 2021)

For example, Bhuiyan et al. have synthesized α-Fe_2O_3 NPs from hexahydrate ferric chloride ($FeCl_3.6H_2O$) using papaya (Carica papaya) leaf extract and obtained ~76.6% RR dye degradation after 6 h (Bhuiyan et al. 2020). Balu et al. have prepared iron oxide Fe_2O_3 NPs by Raphanus sativus leave extract to degrade MB and MR. Both the dyes have undergone complete degradation (100%) in 1 h (Balu et al. 2020). However, functionalization of iron oxide NPs with various ligands such as mercapto-butyric acid or polymers boosts absorption efficiency and removes interference from other metal ions (Aragaw et al. 2021). The majority of iron oxide NCs are amorphous NPs with an average size of about 5 nm, which is a suitable size. These distinctive characteristics of amorphous NPs of iron oxide provide a large surface area for (FeO_x-GO-80) with an iron oxide content of 80 wt% with a prominent mesoporous structure, resulting in increased adsorption sites and, as a result, increased adsorption capacity for the removal of heavy metal pollutants from wastewater (Su et al. 2017; Rashida et al. 2021). A comparative study has been done for dye degradation using various metal oxides NPs and NSs and tabulated in Table 1.

1.1.3 Carbon-based Nanomaterials

Carbon nanomaterials are attractive materials for various applications because of their strength and capability to make bonds with other elements. Various allotropes of

Table 1 Comparison of degradation of dyes by various metal oxide NPs, NSs, and NCs

Nanomaterials	Dye	DC (mg/L)	PC (mg/mL)	Light source	Time (min)	DE (%)	Source
TiO_2 NPs	MO	25	1.5	Sunlight	30–60	90	Ljubas et al. (2015)
TiO_2 NPs	RhB	30	1.5	Microwave irradiation	20	96	Zhong et al. (2009)
N-doped TiO_2	Azo dyes	0.03	0.01	Visible light	240	97	Liu et al. (2005)
C-TiO_2 NPs	RB-19	–	1.6	Visible light	120	100	Helmy et al. (2018)
ZnO NPs	MB	20–100	0.25	UV	180	92.5	Balcha et al. (2016)
ZnO NPs	Rh B	–	–	UV	70	95	Rahman et al. (2013)
Ag-ZnO NPs	MB	0.20		Visible light	180	98.66	Singh et al. (2017)
(Er, Yb)-ZnO NPs	MO	0.004	6	Visible light	90	100	Ahmad (2019)
Cr-ZnS NPs	MO	0.025	5	UV	300	65.22	Eyasu et al. (2013)
ZnS NPs	MR	–	–	Visible	120	95.10	Ye et al. (2018)
Fe_2O_3 NPs	RB-4	20	0.15	UV	56	95.08	Su et al. (2017)
FeO NPs	MG	100	0.4	Sunlight	300	97	Bibi et al. (2019)
ZrO_2/GO	RhB	8	0.5	UV	40	100	Rani et al. (2016)
Au NPs	MB	10	0.15	UV	15	87	Leon et al. (2016)
Nanocupper	MO	20	0.2	Visible	88	35	Liu et al. (2016)
TiO_2/2/β-FeOOH	MO	80	0.2	Visible	120	7.50	Xu et al. (2013)
Ag/Fe_3O_4	MB	40	1	UV	30	99	Liu et al. (2018)

(continued)

Table 1 (continued)

Nanomaterials	Dye	DC (mg/L)	PC (mg/mL)	Light source	Time (min)	DE (%)	Source
CeO_2NCs	MG	30	0.5	UV	21	90	Madhukar Sreekanth et al. (2019)
Ag/COW	RY	100	1	UV	110	96.05	Yola et al. (2014)
Pd/Fe_3O_4-Al	MO	20	0.05	UV	2	90	Cui et al. (2017)
Fe_3O_4/PDA/Ag	MB	40	1	UV	4	100	Cui et al. (2018)
Fe–Ni-PVP NPs	RhB	125	0.75	UV	120	97.44	Kale and Kane (2018)
Pt/N/TiO_2	RB	10	0.3	Visible	90	83.4	Huang et al. (2007)
Ag NPs	MO	10	0.1	Visible	120	51	Jyoti and Singh (2016)
Ag NPs	CR	35.5	0.0002	UV	15	85	Kolya et al. (2015)

carbon like fullerene, graphene, etc., are CNTs (Fig. 6) and are utilized for wastewater treatment (Selvaraj et al. 2020).

Fullerenes

In 1985, Fullerene was discovered and gained considerable attraction because of its remarkable photochemical and photophysical features (Selvaraj et al. 2020). Fullerenes formed a cage-like structure (Fig. 6) with twelve 5-member rings and an unspecified number of 6-member rings. Fullerenes are most commonly found in the form of hexagonal rings with carbon atoms organized inside, but sometimes they also contain pentagonal rings. It is prominent that structures with lesser hexagons show sp^3 bonding, high strain energy. Several research studies have shown that fullerene behaves as an adsorbent to adsorb organic waste and heavy metal ions from wastewater. For example, in wastewater, hydrophobic organic compounds such as naphthalene could be adsorbed by adsorbent C_{60} fullerene, which was coated as a thin film and disseminated in water by magnetic mixing. As a result of its hydrophobic surface, C_{60} fullerene is projected to be an ideal adsorbent for a wide range of organic compounds present in contaminated water (Selvaraj et al. 2020; Geim and Novoselov 2007).

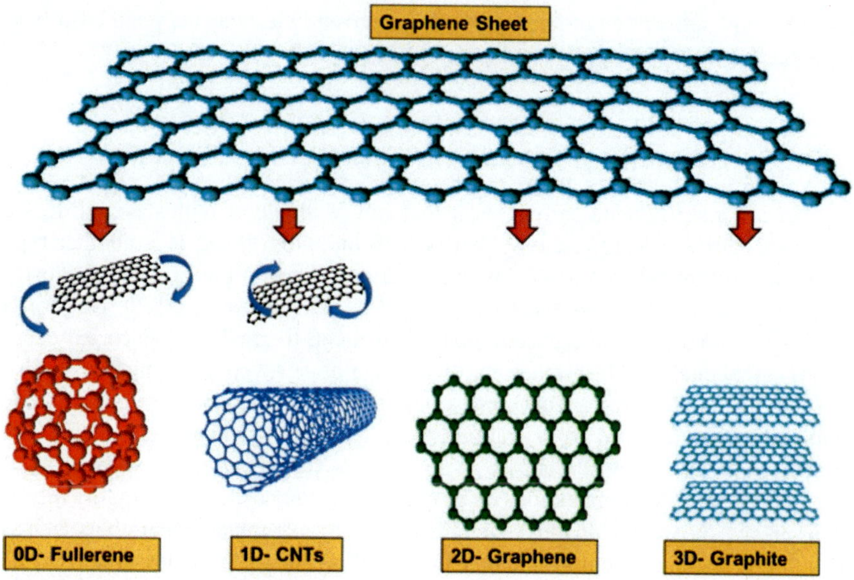

Fig. 6 Graphene as a building block of other forms. An illustration of different allotropes of carbon emerging from a graphene sheet. Redrawn from Geim and Novoselov (2007)

Carbon Nanotubes (CNTs)

CNTs are made up of graphene sheets wrapped into cylinders with diameters ranging from 1 to 100 nm. Their unusual structure and electrical properties make nanotubes attractive for basic research and various applications like the adsorption process (Selvaraj et al. 2020). Its advantages in treating wastewater are due to (i) the ability to adsorb various types of pollutants, (ii) fast adsorption kinetics, (iii) greater specific surface area, and (iv) selectivity to aromatics. CNTs are very effective in eliminating bacterial pathogens. It has been extensively used to remove biological impurities and has received special attention because of its excellent capability to eliminate biological contaminants from wastewater. It has antibacterial properties against various micro-organisms, including bacteria like *Escherichia coli* and Salmonella. Compared with carbon-based adsorbents, the adsorption of cyanobacterial toxins on carbon nanotubes is also greater, mainly because of the larger specific surface area, larger outer diameter, and large mesoporous volume of CNTs. Carbon nanotubes have extraordinary structures and unique characteristics, making them a good candidate for adsorption phenomena, such as metal removal. Single-walled carbon nanotubes (SWCNT) and multi-walled carbon nanotubes (MWCNT) are two types of carbon nanotubes. Both single-walled and multi-walled carbon nanotubes have unique features, such as greater surface area available for adsorption, accessibility of adsorption sites, light mass density. Compared with activated carbon, they also have high adsorption efficiency, which is currently used as the main adsorbent in the purification of water (Rashida et al. 2021).

Graphene

Graphene is a sp^2-bonded carbon sheet that can be single or multi-layered. It is a hexagonal lattice of carbon atoms that is only one atom thick. It is another type of carbon nanomaterial. It has many excellent properties in physics and chemistry (Selvaraj et al. 2020; Geim and Novoselov 2007; Duklan et al. 2020). Due to its distinctive 2D structure and exceptional mechanical, thermal, and electrical capabilities, graphene, which is made up of a few atomic layered graphites, has also been used to analyze the adsorption of pollutants of wastewater. Graphene oxide nanosheets manufactured from graphite using an improved Hummers method are reported to be used as adsorbents to remove CO^{2+}, Cd^{2+} from many aqueous solutions. The adsorption of metal ions on graphene oxide nanosheets was highly reliant on pH and weakly reliant on ionic strength, as seen by manipulating variables such as pH, ionic strength, and humic acid on CO^{2+}, Cd^{2+}. When the pH is less than 100 °C, the presence of humic acid decreases the adsorption of CO^{2+}, Cd^{2+}, on graphene oxide nanosheets. At pH 6.0, the maximum adsorption capacity of CO^{2+}, Cd^{2+} on graphene nanosheets is approximately 106.3 and 68.0 mg/gm, and the temperature is approximately 303 K (Thines et al. 2017). According to this research, graphene is considered the main material for the purification of water if they were manufactured at a larger scale at an affordable cost. Several studies have shown that graphene can be used not only to adsorb heavy metals but also to adsorb fiber dyes for the purification of water (Das et al. 2020). It has been reported that graphite is used for adsorption of dye from an aqueous solution after oxidizing the graphite by using the Hummers-Offeman method (Geim and Novoselov 2007).

2 Removal Techniques

Hazardous pollutants like heavy metals, organic pollutants, and anions are introduced into the system of freshwater supply through industrial and agricultural waste. The types of new organic pollutants cannot be deteriorated by the chemical, biological, and photolytic processes in the environment, including pesticides, drugs, hormones, types of various aromatic compounds, etc. (Thines et al. 2017). The traditional methods for purification of wastewater cannot remove all the contaminants, and even the very low concentration of pollutants will cause to form dangerous disinfection by-products (DBP). The presence of contaminants affects health and hygiene and reduces food safety. The availability of clean water is necessary for food production, including preparation, distribution, and consumption. Due to the increase in the number of water-borne diseases caused by the number of inorganics, organic hazardous waste, the new and innovation of effective treatment processes are vital. Nanotechnology has shown excellent results in the removal of the aforementioned type of pollutants. Many studies have been conducted on wastewater treatment processes supported by nanotechnology, many of which have shown superior performance over traditional technologies (Figoli et al. 2017b). According to the types of nanomaterials, wastewater treatment is divided into three categories: (i) nanoadsorption, (ii) nanocatalyst, and (iii) nanomembrane.

2.1 *Nanoadsorption*

Adsorption is a surface phenomenon in which contaminants are adsorbed onto a solid surface. Adsorption occurs in all physical forces, but it can also be linked to weak chemical bonding in specific cases. Nanoadsorption is generally used in wastewater treatment to remove organic and inorganic pollutants. The distinctive characteristics of the nanoadsorbent, such as small size, high catalytic potential, high reaction activity, high surface area, ease of separation, and a high number of active sites for interaction with various pollutants, are helpful for wastewater treatment (Table 2).

Table 2 Comparison of maximum adsorption capacity of different NCs against organic dyes

Adsorbent	Pollutant	Dye volume (mg/L)	Adsorbent volume (g/L)	MAC mg/g	Ref.
Ag NPs	IC	3.55	0.4	73.05	Gemeay et al. (2018)
RGO-CNT-PPD	MO	30	0.1	294	Sarkar et al. (2014)
Polypyrrole	MO	50	0.8	143.89	Alghandi et al. (2019)
Banana peel	Rh. B	100	0.5	28.8	Akter et al. (2021)
NZVI	MB	10	0.5	208.33	Arabi and Reza Sohrabi (2014)
GO	MG	50	0.2	416.7	Sykam et al. (2018)
Au-RGO	MB	15	0.25	338.65	Dutta et al. (2013)
Ag NPs	MB	50	1	213.7	Karthiga Devi et al. (2016)
MPA/PMNPs	CV	25	0.5	88.65	Ali et al. (2018b)
rGO/PVA	MB	20	4	231.12	Cheng et al. (2015)
Au NPs/AC	MO	20	0.005	161.29	Ghaedi et al. (2015)

2.2 Carbon-based Nanoadsorbents

Carbon-based nanoadsorbents are, such as graphene, graphene oxide, and carbon nanotubes. Carbon nanotubes can be divided into single-walled nanotubes (SWNT) and multi-walled nanotubes (MWNT) as advanced water purifiers. These materials are rarely used in their pure form. Still, the most common composition, either dispersed in a polymer or decorated with metal NPs, such as silver (Ag) and iron oxide (Fe_3O_4) NPs (Smith and Rodrigues 2015). Due to their hydrophobic surface, the carbon nanotubes dissolve the bundle in an aqueous medium, which reduces the active surface area. These aggregates are high-energy locations for adsorbing organic pollutants in the water. The reason for the adsorption is as follows: (i) the availability of larger pores in carbon nanotubes bundles and (ii) greater accessible adsorption sites.

2.3 Metal-based Nanoadsorbents

In the water purification process, metal-based nanoadsorbents, like zinc oxide, titanium oxide, iron oxide, are employed to remove heavy metals. These adsorbents are both efficient and cost-effective. The oxygen in metallic oxides forms a compound with the heavy metals found in wastewater. That is how they work. Magnetic nanoadsorbents, for example, maghemite (γ Fe_2O_3), hematite (α-Fe_2O_3), and spinel ferrite ($M^{2+}Fe_2O_4$, Where M^{2+}: Co^{2+}, Fe^{2+}, etc.) are very excellent adsorption materials used to collect and eliminate toxic and carcinogenic pollutants from wastewater (Ahmad et al. 2021). The environmental advantage is reflected in its magnetism. They can be easily detached from the reaction medium by applying an external magnetic field. Various studies proved that metal-based nanoadsorbents remove various elements from wastewater, for example, ionic forms of lead, nickel, arsenic, chromium, cobalt, etc. ZnO nanoadsorbents are employed for the removal of Zn^{2+}, Cd^{2+}, and Hg^{2+} ions from aqueous solutions in work published in the literature. Metals ions are adsorbed onto ZnO NPs at discrete amounts. Because Hg^{2+} has the smallest hydrated ionic radius among metal ions, it has the highest adsorption capacity (Ahmad et al. 2021). Alumina nanoadsorbents can also be manufactured because of their large surface area, good thermal stability, and relatively inexpensive. They are used to remove metal ions, like cadmium, chromium, copper, lead, and mercury. Metal-based nanoadsorbents are employed for the elimination of heavy metals in water and high-efficiency nanoadsorbents in wastewater because of their merits, like large adsorption capacity, faster kinetics, etc. (Ahmad et al. 2021).

2.4 Polymer-based Nanoadsorbents

Polymer-based nanoadsorbents have recently acquired attraction toward wastewater treatment. They are used as a structure that can insert nanoscale inorganic materials or a bed or template for preparing NPs. Polymer-based magnetic nanoparticles efficiently removed heavy metal ions like Zn^{2+}, Cd^{2+}, Pb^{2+} from aqueous solutions and have a high maximum capacity of adsorption of pH 5.5 (Ahmad et al. 2021; Kumar et al. 2011). This nanoadsorbent can be reused for at least four cycles. In recent studies, bi-metal micro- and nano-multifunctional polymeric adsorbents were developed for the elimination of fluoride and arsenic (V). Suspension polymerization is used to make the polymer. To make bi-metal-doped nanoadsorbents, aluminum and iron salts are added during the polymerization process. When compared to fluoride, iron-doped nanoadsorbents had excellent adsorption for arsenic, while aluminum-doped nanoadsorbents had excellent adsorption for fluoride (Kumar et al. 2011). Hence, polymer-based nanoadsorbents are magnificent materials due to their structures, pore sizes, and tunable functional groups; making them selective for a specific contaminant is challenging to remove heavy metal ions from wastewater. The adsorptive capacity is quite low, and regeneration is required when CO is high.

2.5 Nanocatalysis

Nanocatalysis is a rapidly emerging technology in which nanomaterials are used as catalysts in various applications such as reduced global warming, wastewater treatment. Different types of NPs are used as the catalyst for eliminating organic contaminants such as pesticides, dyes, fertilizers, oil grease and inorganic pollutants such as calcium, potassium, chloride, sulfate, nitrate. Nowadays, NPs of titanium oxide (TiO_2) have emerged as an attractive photocatalyst for water treatment (Adesina 2004). TiO_2 is highly adaptable; they can be used in various applications and can act as an oxidizing and reducing catalyst for the removal of organic and inorganic contaminants from wastewater. The addition of NPs of TiO_2 substantially improved the deterioration of organic pollutants in wastewater in the presence of ultraviolet radiation (Fig. 7) (Nawaz et al. 2020). It is reported that NPs of TiO_2 effectively (i) deteriorate organic pollutants like chlorinated alkanes and benzene, dioxins, furans, etc. and (ii) remove toxic metal ions like Cr^{6+}, Pt^{2+} in aqueous solutions under ultraviolet light. When ultraviolet light in the range of 200–390 nm is irradiated on TiO_2, electron–hole pairs are photoexcited. They move to the conduction and valence bands, leading to the separation of charge for an efficient photocatalytic function that depends on the substrate's redox potential. As a result, in a pretreatment phase, the biodegradability of decomposable elements can be improved. Primarily, steady mixtures, like anti-microbial or other micro-contaminants, might be removed by photocatalysis polishing (Ahmad et al. 2021).

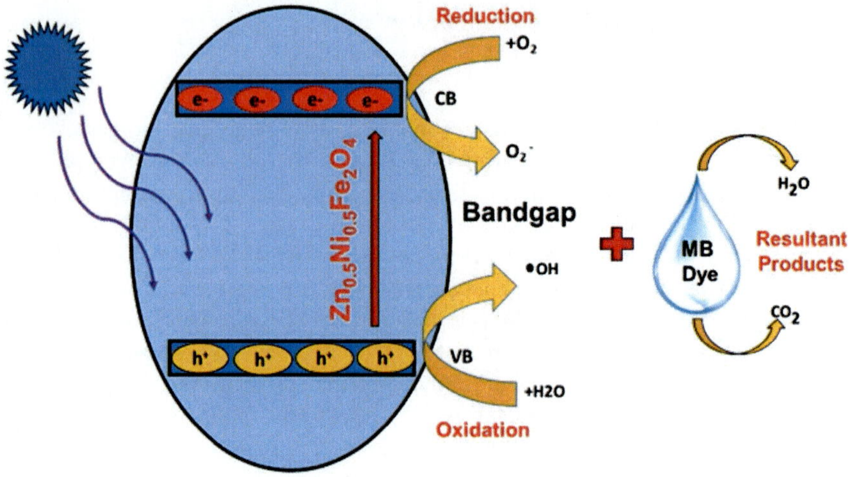

Fig. 7 Schematic of degradation of MB organic dyes. Redrawn and reprinted with permission from Nawaz et al. (2020)

TiO_2 is activated by ultraviolet light, but daylight or apparent light lamps are additionally allowed. KRONO clean 7000, a photocatalyst bandgap moved toward smaller energy: This contributes toward using a broader spectrum in sunlight (Gehrke et al. 2015). Modified technology has been explored to enhance the photocatalytic performance of titanium dioxide, including activity-enhancing or redshift for saving of energy, such as the combination of nano-silica (good thermal and chemical stability) and nano-titanium dioxide (nano-semiconductor) creates new surface-active sites. The catalytic performance of the silica/titanium dioxide nanocomposite is highly dependent on the content and distribution of TiO_2. Photocatalysis has a promising future as a long-term, eco-friendly, and cost-effective water purification technique. However, there are several technical hurdles to overcome before it can be used on a wider scale, for example, (i) catalyst tuning to enhance quantum yield or to use visible light, (ii) designing an effective photocatalytic reactor, and (iii) upgrade reaction selectivity.

2.6 *Membranes for Water Purification*

A nanomembrane is a permeable thin-layered membrane having pores sizes of 1–10 nm that enables water molecules to pass through it while preventing bacteria, viruses, heavy metals, pesticides, etc., from passing through them. The membrane's operation is dependent whether on pressure-driven or electrical technology. Pressure-driven membrane technique is an excellent approach for wastewater treatment (Kumar et al. 2014). Membrane filtration procedures have become more sophisticated techniques of industrial wastewater treatment. Membranes segregate materials based on the size of the pore and the molecule. It is a sustainable and systematic method for the purification of wastewater. The membrane material determines the membrane system's efficacy. Membrane permeability, theoretical resistance, thermal and mechanical stabilities are enhanced by incorporating functional nanomaterials into membranes. Surface-functionalized membranes and nanocomposite membranes, which can be produced from mixed materials, are practical filtration units. Nanofillers are used in mixed matrix membranes, and the majority of them are inorganic. They have a large surface area and are incorporated into polymeric or inorganic oxide matrix (Sarkar et al. 2014). Hydrophobic membranes (Fig. 8) are used in various industries for industrial wastewater treatment. For these various hydrophilic metal oxides, nanomaterials are used, such as Al_2O_3, TiO_2, and zeolites. Ag NPs, CNTs, bimetallic NPs are also used for membrane filtration (Karthiga Devi et al. 2016).

NCs membranes are made up of nanofillers, which are thin-film polymeric grids made up of a sequential arrangement of mesoporous carbons. They are semi-permeable, and reverse osmosis uses the top surface. Hydrophobic mesoporous carbons are transformed to hydrophilic carbons by atmospheric pressure plasma. A small amount of hydrophilic carbon increases hydrophilicity; this leads to an increase in the permeability of pure water. Thin-film nanocomposites were composed of polyamide and nano-NaX zeolite of 40–150 nm and are covered by interfacial polymerization using trimethyl chloride and m-phenylenediamine monomers over polyethersulfone. This membrane has a high permeability to purify freshwater,

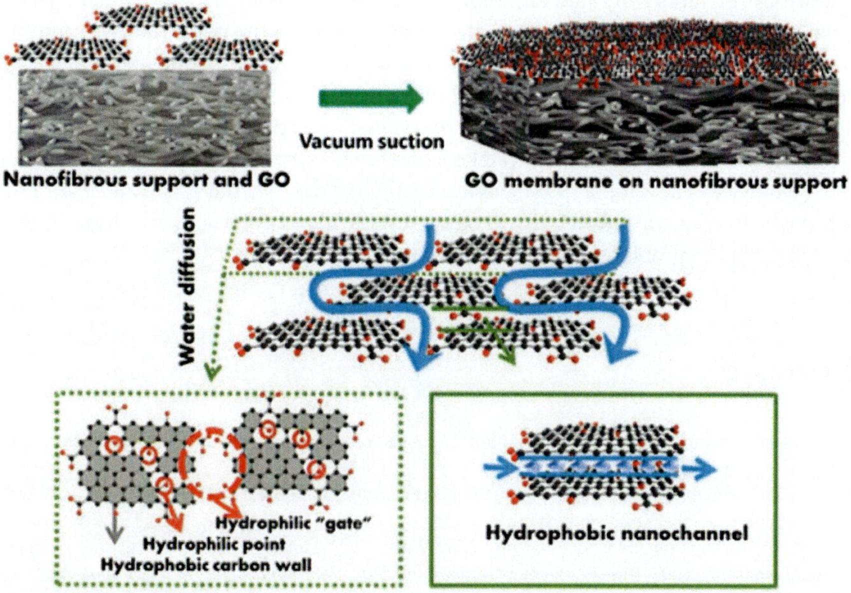

Fig. 8 GO-based porous nanofibrous membrane for water treatment. Reproduced with permission from Wang et al. (2016)

leaving pollutants behind the membrane (Fathizadeh et al. 2011). Electrospinning is a useful strategy for modifying the surface characteristics of nanomaterials, and various nanofibers have been successfully employed for wastewater treatment. These nanofibers have a large surface area and porosity, resulting in a nanofiber mat with a complex pore structure. Nanofibers are highly active against water pathogens, have low toxicity, and minimize health hazards. It is extremely simple to dope functional nanomaterials to form a filtration membrane, which has greater reactivity and selectivity to various pollutants.

3 Conclusion

In the present era, wastewater treatment techniques that would provide high-quality freshwater, eliminate organic and inorganic pollutants, and enhance industrial activities are very important. The opportunity is provided by nanotechnology; the distinctive characteristics of nanoparticles, such as greater surface area, size, shape, and dimensions, make them an excellent choice for water purification. Nanoparticles can be used to remove metal ions, anions, organic chemicals, and micro-organisms. Because the nanoparticle doses required for water purification are minimal, their use is reasonably cost-effective. This chapter highlights various nanotechnologies such as nanoadsorption, nanocatalysis, and nanomembranes, among others. Under UV and solar irradiation, photocatalytic processes successfully remove various types of water impurities, including organic and inorganic pollutants. As reactive oxygen species have a limited lifetime, surface modifications may boost the photocatalytic activity of selected compounds and enhance the affinity of modified nanomaterials toward several rising water pollutants. Bimetallic nanomaterials have also been proven to be useful in the treatment of wastewater contaminants. The prospects of NPs in water treatment are promising, but it will take a combined effort from scientific and corporate resources to develop a rapid, environmentally friendly, and practical system for the purification of wastewater. It will be achievable if everyone works with each other to overcome the problem of worldwide water contamination.

References

A.A. Adesina, Industrial exploitation of photocatalysis: progress, perspectives and prospects. Catal. Surv. Asia **8**(4), 265–273 (2004)

I. Ahmad, Inexpensive and quick photocatalytic activity of rare earth (Er, Yb) co-doped ZnO nanoparticles for degradation of methyl orange dye. Sep. Purif. Technol. **227**, 115726 (2019)

N.A. Ahmad, P.S. Goh, A.K. Zulhairun, T.W. Wong, A.F. Ismail, Chapter 1: the role of functional nanomaterials for wastewater remediation, in *Functional Hybrid Nanomaterials for Environmental Remediation* (2021), pp. 1–28

M. Akter, F.B.A. Rahman, M.Z. Abedin, S.M.F. Kabir, Adsorption characteristics of banana peel in the removal of dyes from textile effluent. Textiles **1**(2), 361–375 (2021)

A.A. Alghandi, A.B. Al-Odayni, W.S. Saeed, M.S. Almutairi, Adsorption of azo dye methyl orange from aqueous solutions using alkali-activated polypyrrole-based graphene oxide. J. MDPI **24**, 3685 (2019)

I. Ali, M. Suhail, Z.A. Alothman, A. Alwarthan, Recent advances in syntheses, properties and applications of TiO_2 nanostructures. RSC Adv. **8**, 30125–30147 (2018a)

I. Ali, C. Peng, Z.M. Khan, M. Sultan, I. Naz, Green synthesis of phytogenic magnetic nanoparticles and their applications in the adsorptive removal of crystal violet from aqueous solution. Arab. J. Sci. Eng. **43**, 6245–6259 (2018b)

M.T. Amin, A.A. Alazba, U. Manzoor, A review of removal of pollutants from water/wastewater using different types of nanomaterials. Adv. Mater. Sci. Eng. **2014**, 825910 (2014)

S. Arabi, M. Reza Sohrabi, Removal of methylene blue, a basic dye, from aqueous solutions using nano-zerovalent iron. J. Water Sci. Tech. **70**(1), 24–31 (2014)

T.A. Aragaw, F.M. Bogale, B.A. Aragaw, Iron-based nanoparticles in wastewater treatment: a review on synthesis methods, applications, and removal mechanisms. J. Saudi Chem. Soc. **25**(8), 101280 (2021)

A. Balcha, O.P. Yadav, T. Dey, Photocatalytic degradation of methylene blue dye by zinc oxide nanoparticles obtained from precipitation and sol-gel methods. Environ. Sci. Pollut. Res. **23**(24), 25485–25493 (2016)

P. Balu, I.V. Asharani, D. Thirumalai, Catalytic degradation of hazardous textile dyes by iron oxide nanoparticles prepared from Raphanus sativus leaves' extract: a greener approach. J Mater Sci. Mater Electron **31**, 10669–10676 (2020)

Md.S.H. Bhuiyan, M.Y. Miah, S.C. Paul, T.D. Aka, O. Saha, M.M. Rahaman, M.J.I. Sharif, O. Habiba, M. Ashaduzzamane, Green synthesis of iron oxide nanoparticle using Carica papaya leaf extract: application for photocatalytic degradation of remazol yellow RR dye and antibacterial activity. Heliyon, **6**(8), e04603 (2020)

I. Bibi, N. Nazar, S. Ata, M. Sultan, A. Ali, A. Abbas, M. Iqbal, Green synthesis of iron oxide nanoparticles using pomegranate seeds extract and photocatalytic activity evaluation for the degradation of textile dye. J. Market. Res. **8**(6), 6115–6124 (2019)

H. Borji, G.M. Ayoub, R. Bilbeisi, N., Nassar, L. Malaeb, How effective are nanomaterials for the removal of heavy metals from water and wastewater? Water Air Soil Pollut. **231**(7), 1–35 (2020)

P. Chamoli, M.K. Das, K.K. Kar, Green synthesis of silver-graphene nanocomposite-based transparent conducting film. Physica E **90**, 76–84 (2017)

P. Chamoli, R.K. Shukla, A.N. Bezbaruah, K.K. Kar, K.K. Raina, Rapid microwave growth of mesoporous TiO_2 nano-tripods for efficient photocatalysis and adsorption. J. Appl. Phys. **130**, 164901 (2021a)

P. Chamoli, R.K. Shukla, A. Bezbaruah, K.K. Kar, K.K. Raina, Microwave-assisted rapid synthesis of honeycomb core-ZnO tetrapods nanocomposites for excellent photocatalytic activity against different organic dyes. Appl. Surf. Sci. **555**, 149663 (2021b)

Z. Cheng, J. Liao, B. He, F. Jhang, X. Huang, L. Zhou, One-step fabrication of graphene oxide enhanced magnetic composite gel for highly efficient dye adsorption and catalysis. J. ACS Sustain. Chem. Eng. **3**(7), 1677–1685 (2015)

B.L. Cínti, Z.F. PolonioJoão, A. Pamphile, J.C. Polonio, Effects of textile dyes on health and the environment and bioremediation potential of living organisms. Biotechnol. Res. Innov. **3**(2), 275–290 (2019)

X. Cui, Y. Zheng, M. Tian, Z. Dong, Novel yolk–shell-structured $Fe_3O_4@\gamma$-AlOOH nanocomposite modified with Pd nanoparticles as a recyclable catalyst with excellent catalytic activity. J. Appl. Sci. Surf. **416**, 103–111 (2017)

K. Cui, B. Yan, Y. Xie, H. Qian, X. Wang, Q. Huang, Y. He, S. Jin, H. Zeng, Regenerable urchin-like $Fe_3O_4@$ PDA-Ag hollow microspheres as catalyst and adsorbent for enhanced removal of organic dyes. J. Hazard. Mat. **350**, 66–75 (2018)

T.K. Das, T.S. Sakthivel, A. Jeyaranjan, S. Seal, A.N. Bezbaruaha, Ultra-high arsenic adsorption by graphene oxide iron nanohybrid: removal mechanisms and potential applications. Chemosphere **253**, 126702 (2020)

N. Duklan, P. Chamoli, K.K. Raina, R.K. Shukla, Dye dispersed lyotropic liquid crystals: soft materials with high ionic conductivity and self-sustained adsorbents for dye sequestration. Inorg. Chem. Commun. **116**, 107924 (2020)

S. Dutta, S. Sarkar, C. Ray, T. Pal, Benzoin derived reduced graphene oxide (rGO) and its nanocomposite: application in dye removal and peroxidase-like activity. RSC Adv. **3**(44), 21475–21483 (2013)

A. Eyasu, O.P. Yadav, R.K. Bachheti, Photocatalytic degradation of methyl orange dye using Cr-doped ZnS nanoparticles under visible radiation. Int. J. Chem. Tech. Res. **5**(4), 1452–1461 (2013)

F. Fan, X. Wang, Y. Ma, K. Fu, Y. Yang, Enhanced photocatalytic degradation of dye wastewater using ZnO/reduced graphene oxide hybrids. Fuller. Nanotub. Car. Nanostruct. **23**, 917–921 (2015)

M. Fathizadeh, A. Aroujalian, A. Raisi, Effect of added NaX nano-zeolite into polyamide as a top thin layer of membrane on water flux and salt rejection in a reverse osmosis process. J. Membr. Sci. **375**(1–2), 88–95 (2011)

A. Figoli, M.S.S. Dorraji, A.R. Amani-Ghadim, *Application of nanotechnology in drinking water purification* (Academic Press, Water purification, 2017b), pp. 119–167

A. Figoli, M.S.S. Dorraji, A.R. Amani-Ghadim, Application of nanotechnology in drinking water purification, in *Water Purification* (Academic Press, 2017a), pp. 119–167

A. Gautama, A. Kshirsagara, R. Biswasa, S. Banerjee, P.K. Khann, Photodegradation of organic dyes based on anatase and rutile TiO_2 nano-particles. RSC Adv. **6**, 2746–2759 (2016)

I. Gehrke, A. Geiser, A. Somborn-Schulz, Innovations in nanotechnology for water treatment. Nanotechnol. Sci. Appl. **8**, 1–17 (2015)

A.K. Geim, K.S. Novoselov, The rise of graphene. Nat. Mater. **6**, 183–191 (2007)

A.H. Gemeay, E.F. Aboelfetoh, R.G. El-Sharkawy, Immobilization of green synthesized silver nanoparticles onto amino-functionalized silica and their application for indigo carmine dye removal. Water Air Soil Pollut. **229**, 16 (2018)

M. Ghaedi, F. Mohammdi, A. Ansari, Gold nanoparticles loaded on activated carbon as novel adsorbent for kinetic and isotherm studies of methyl orange and sunset yellow adsorption. J. Dispersion Sci. Tech. **36**, 652–659 (2015)

D. Gola, A. kritia, N. Bhatt, M. Bajpai, A. Singh, A. Arya, N. Chauhan, S.K. Srivastava, P.K. Tyagi, Y. Agrawal, Silver nanoparticles for enhanced dye degradation. Curr. Res. Green Sustain. Chem. **4**, 100132 (2021)

J.E.S. Haggerty, L.T. Schelhas, D.A., Kitchaev, High-fraction brookite films from amorphous precursors. Sci. Rep. **7**, 15232 (2017)

E.T. Helmy, A. El Nemr, M. Mousa, E. Arafa, S. Eldafrawy, Photocatalytic degradation of organic dyes pollutants in the industrial textile wastewater by using synthesized TiO_2, C-doped TiO_2, S-doped TiO_2 and C, S co-doped TiO_2 nanoparticles. J. Water Environ. Nanotechnol. **3**(2), 116–127 (2018)

L.H. Huang, C. Sun, Y.L. Liu, Pt/N-codoped TiO_2 nanotubes and its photocatalytic activity under visible light. Appl. Surf. Sci. **253**, 7029–7035 (2007)

M. Jabeen, M. Ishaq, W. Songz, L. Xu, I. Maqsood, Q. Deng, UV-assisted photocatalytic synthesis of ZnO-reduced graphene oxide nanocomposites with enhanced photocatalytic performance in degradation of methylene blue. J. Solid State Sci. Technol. **6**, 36–43 (2017)

A. Jain, F. Ahmad, D. Gola, A. Malik, N. Chauhan, P. Deye, P.K. Tyagic, Multi dye degradation and antibacterial potential of Papaya leaf derived silver nanoparticles. Environ. Nanotechnol. Monit. Manage. **14**, 100337 (2020)

K. Jyoti, A. Singh, Green synthesis of nanostructured silver particles and their catalytic application in dye degradation. J. Genet. Eng. Biotech. **14**(2), 311–317 (2016)

R.D. Kale, P.B. Kane, Decolourization by PVP stabilized Fe-Ni nanoparticles of reactive black 5 dye. J. Environ. Chem. Eng. **6**(5), 5961–5969 (2018)

G. Karthiga Devi, P. Senthil kumar, K. Sathish Kumar, Green synthesis of novel silver nanocomposite hydrogel based on sodium alginate as an efficient biosorbent for the dye wastewater treatment: prediction of isotherm and kinetic parameters. J. Desalin. Water Treat. **57**, 27686–27699 (2016)

G. Kaur, M. Kaur, A. Thakur, A. Kumar, FeS_2 pyrite nanostructures: an efficient performer in photocatalysis, in *Green Methods for Wastewater Treatment*, ed. by M. Naushad, S. Rajendran, E. Lichtfouse (Springer Nature Switzerland AG, 2020), p. 55

S.T. Khan, A. Malik, Engineered nanomaterials for water decontamination and purification: from lab to products. J. Hazard. Mater. **363**, 295–308 (2019)

H. Kolya, P. Maiti, A. Pandey, T. Tripathy, Green synthesis of silver nanoparticles with antimicrobial and azo dye (Congo red) degradation properties using Amaranthus gangeticus Linn leaf extract. J. Anal. Sci. Tech. **6**, 33–40 (2015)

S. Kumar, W. Ahlawat, G. Bhanjana, S. Heydarifard, M.M. Nazhad, N. Dilbaghi, Nanotechnology-based water treatment strategies. J. Nanosci. Nanotechnol. **14**(2), 1838–1858 (2014)

V. Kumar, N. Talreja, D. Deva, N. Sankararamakrishnan, A. Sharma, N. Verma, Development of bi-metal doped micro-and nano multi-functional polymeric adsorbents for the removal of fluoride and arsenic (V) from wastewater. Desalination **282**, 27–38 (2011)

N. Kumaresan, M.M.A. Sinthiya, K. Ramamurthi, R.R. Babu, K. Sethuraman, Visible light driven photocatalytic activity of ZnO/CuO nanocomposites coupled with rGO heterostructures synthesized by solid-state method for RhB dye degradation. Arab. J. Chem **13**(2), 3910–3928 (2020)

K.R. Kunduru, M. Nazarkovsky, S. Farah, R.P. Pawar, A. Basu, A.J. Domb, Nanotechnology for water purification: applications of nanotechnology methods in wastewater treatment, in *Water Purification* (2017), pp. 33–74

G.J. Lee, J.J. Wu, Recent developments in ZnS photocatalysts from synthesis to photocatalytic applications—a review. Powder Technol. **318**, 8–22 (2017)

E.R. Leon, E.L. Rodriguez, C.R. Beas, G.P. Villa, R.A.I. Palomares, Study of methylene blue degradation by gold nanoparticles synthesized within natural zeolites. J. Nanomater. 1–10 (2016)

Y. Liu, X. Chen, J. Li, C. Burda, Photocatalytic degradation of azo dyes by nitrogen-doped TiO_2 nanocatalysts. Chemosphere **61**(1), 11–18 (2005)

D. Liu, G. Wong, J. Lin, Y. He, X. Li, Z. Li, Photocatalysis using zero-valent nano-copper for degrading methyl orange under visible light irradiation. Opti. Mate. **53**, 155–159 (2016)

Y. Liu, Y. Zhang, Q. Kou, Y. Chen, D. Han, D. Wang, Z. Lu, L. Chen, J. Yang, S. Xing, Eco-friendly seeded Fe_3O_4-Ag nanocrystals: a new type of highly efficient and low-cost catalyst for methylene blue reduction. RSC Ad. **8**, 2209–2218 (2018)

D. Ljubas, G. Smoljanić, H. Juretić, Degradation of methyl orange and Congo red dyes by using TiO_2 nanoparticles activated by the solar and the solar-like radiation. J. Environ. Manage. **161**, 83–91 (2015)

H. Lu, J. Wang, M. Stoller, T. Wang, Y. Bao, H. Hao, An overview of nanomaterials for water and wastewater treatment. Adv. Mater. Sci. Eng. **2016** (2016)

T.V. Madhukar Sreekanth, P.C. Nagajyothi, G.R. Reddy, J. Shim, K. Yoo, Urea assisted ceria nanocubes for efficient removal of malachite green organic dye from aqueous system. Sci. Rep. **9**, 14477 (2019)

S.K. Maji, A.K. Dutta, D.N. Srivastava, P. Paul, A. Mondal, B. Adhikarya, Effective photocatalytic degradation of organic pollutant by ZnS nanocrystals synthesized via thermal decomposition of single-source precursor. Polyhedron **30**(15), 2493–2498 (2011)

M. Maninder, X. Baojun, A critical review on anti-diabetic and anti-obesity effects of dietary resistant starch. Crit. Rev. Food Sci. Nutr. **59**(18), 3019–3031 (2019)

M. Nasrollahzadeh, M. Sajjadi, S. Iravani, R.S. Varma, Carbon-based sustainable nanomaterials for water treatment: state-of-art and future perspectives. Chemosphere **263**, 128005 (2021)

A. Nawaz, A. Khan, N. Ali, N. Ali, M. Bilal, Fabrication and characterization of new ternary ferrites-chitosan nanocomposite for solar-light driven photocatalytic degradation of a model textile dye. Environ. Technol. Innov. **20**, 101079 (2020)

Ü. Özgür, Ya.I. Alivov, C. Liu, A. Teke, M.A. Reshchikov, S. Doğan, V. Avrutin, S.-J. Cho, H. Morkoç, A comprehensive review of ZnO materials and devices. J. Appl. Phys. **98**, 041301 (2005)

S. Pandey, J.Y. Do, J. Kim, M. Kanga, Fast and highly efficient catalytic degradation of dyes using κ-carrageenan stabilized silver nanoparticles nanocatalyst. Carbohydr. Polym. **230**(15), 115597 (2020)

Q.I. Rahman, M. Ahmad, S.K. Misra, M. Lohani, Effective photocatalytic degradation of rhodamine B dye by ZnO nanoparticles. Mater. Lett. **91**, 170–174 (2013)

R. Rajkumar, G. Ezhumalai, M. Gnanadesigan, A green approach for the synthesis of silver nanoparticles by Chlorella vulgaris and its application in photocatalytic dye degradation activity. Environ. Technol. Innov. **21**, 101282 (2021)

S. Rani, M. Aggarwal, M. Kumar, S. Sharma, D. Kumar, Removal of methylene blue and rhodamine B from water by zirconium oxide/graphene. J. Water Sci. **30**, 51–60 (2016)

U.S. Rashida, T.K. Dasa, T.S. Sakthivel, S. Seal, A.N. Bezbaruaha, GO-CeO_2 nanohybrid for ultra-rapid fluoride removal from drinking water. Sci. Total Environ. **793**, 148547 (2021)

K. Ravi, B.S. Mohan, G.S. Sree, I.M. Raju, K. Basavaiah, B.V. Rao, ZnO/RGO nanocomposite via hydrothermal route for photocatalytic degradation of dyes in presence of visible light. Int. J. Chem. Stud. **6**, 20–26 (2018)

C. Sarkar, C. Bora. S.K. Dolui, Selective dye adsorption by pH modulation on amine-functionalized reduced graphene oxide–carbon nanotube hybrid. J. Ind. Eng. Chem. Res. **53**(42), 16148–16155 (2014)

K. Sathiyana, R. Bar-Zivb, O. Mendelson, T. Zidkia, Controllable synthesis of TiO_2 nanoparticles and their photocatalytic activity in dye degradation. Mater. Res. Bull. **126**, 110842 (2020)

M. Selvaraj, A. Hai, F. Banata, M.A. Haija, Application and prospects of carbon nanostructured materials in water treatment: a review. J. Water Process Eng. **33**, 100996 (2020)

M. Sharma, T. Jain, S. Singh, O.P. Pandey, Photocatalytic degradation of organic dyes under UV–visible light using capped ZnS nanoparticles. Sol. Energy **86**(1), 626–633 (2012)

R. Singh, P.B. Barman, D. Sharma, Synthesis, structural and optical properties of Ag doped ZnO nanoparticles with enhanced photocatalytic properties by photo degradation of organic dyes. J. Mater. Sci. Mater. Electron. **28**(8), 5705–5717 (2017)

S.C. Smith, D.F. Rodrigues, Carbon-based nanomaterials for removal of chemical and biological contaminants from water: a review of mechanisms and applications. Carbon **91**, 122–143 (2015)

H. Su, Z. Ye, N. Hmidi, High-performance iron oxide–graphene oxide nanocomposite adsorbents for arsenic removal. Colloids Surf. A **522**, 161–172 (2017)

N. Sykam, V. Madhavi, G. Mohan Rao, Rapid and efficient green reduction of graphene oxide for outstanding supercapacitors and dye adsorption applications. J. Environ. Chem. Eng. **6**(2), 3223–3232 (2018)

A.M. Tayeb, D.S. Hussein, Synthesis of TiO_2 nanoparticles and their photocatalytic activity for methylene blue. Am. J. Nanomater. **3**, 57–63 (2015)

R.K. Thines, N.M. Mubarak, S. Nizamuddin, J.N. Sahu, E.C. Abdullah, P. Ganesan, Application potential of carbon nanomaterials in water and wastewater treatment: a review. J. Taiwan Inst. Chem. Eng. **72**, 116–133 (2017)

J. Wang, P. Zhang, B. Liang, Y. Liu, T. Xu, L. Wang, B. Cao, K. Pan, Graphene oxide as an effective barrier on a porous nanofibrous membrane for water treatment. ACS Appl Mater. Interfaces **8**(9), 6211–6218 (2016)

P. Xu, G.M. Zeng, D.L. Huang, C.L. Feng, S. Hu, M.H. Zhao, C. Lai, Z. Wei, C. Huang, G.X. Xie, Z.F. Liu, Use of iron oxide nanomaterials in wastewater treatment: a review. Sci. Total Environ. **424**, 1–10 (2012)

Z. Xu, M. Zhang, J. Wu, J. Liang, L. Zhou, B. Lu, Visible light-degradation of azo dye methyl orange using TiO 2/β-FeOOH as a heterogeneous photo-Fenton-like catalyst. Water Sci. Tech. **68**(10), 2178–2185 (2013)

Z. Ye, L. Kong, F. Chen, Z. Chen, Y. Lin, C. Liu, A comparative study of photocatalytic activity of ZnS photocatalyst for degradation of various dyes. Optik **164**, 345–354 (2018)

M.L. Yola, T. Eren, N. Atar, S. Wang, Adsorptive and photocatalytic removal of reactive dyes by silver nanoparticle-colemanite ore waste. J. Chem. Eng. **242**, 333–340 (2014)

S. Zhang, Preparation of controlled-shape ZnS microcrystals and photocatalytic property. Ceram. Int. **40**(3), 4553–4557 (2014)

H.E. Zhong, Y.A.N.G. Shaogui, J.U. Yongming, S.U.N. Cheng, Microwave photocatalytic degradation of Rhodamine B using TiO_2 supported on activated carbon: mechanism implication. J. Environ. Sci. **21**(2), 268–272 (2009)

Nanoparticles and Their Role in Environmental Decontamination Technologies

M. P. Ajith and Rajamani Paulraj

Abstract Environmental contamination is a severe problem for both developing and developed nations. Due to anthropogenic activities, the pollution load gets accumulated in the environment. Both organic and inorganic contaminants are harmful to the living being. Their incidence and persistence have increased substantially in recent years. For the remediation of impurities from diverse ecological media such as water, soil, and air, ecological decontamination technologies primarily rely on numerous technologies such as adsorption, absorption, chemical reactions, filtration, and photocatalysis. Nanotechnology has recently gained popularity in water, soil, and air decontamination. Detection, decontamination, and pollution preclusion are the primary elements of nanoparticle-based remediation. Nanotechnology has been extensively used in numerous fields, including environmental remediation. This chapter briefly explains the recent advancement of nanotechnology and nanomaterials in environmental decontamination technologies.

Keywords Decontamination · Nanotechnology · Organic nanoparticles · Metal and metal oxide nanoparticles · Polymer-based nanoparticles

1 Introduction

Our environment has become polluted with different sorts of pollutants. Solid, liquid, and gaseous pollutants are among the toxins created by human activities for short-term economic gain at the expenditure of long-term environmental gains for humankind. A diversity of pollutants such as heavy metals, pesticides, pharmaceutical chemicals, polyaromatic hydrocarbons, organic solvents, and inorganic solvents contaminates water, land, and air, posing a hazard to the ecosystem and human health. Pollutants end up in the environment as a result of industrial and commercial operations, oil and chemical spills, and non-point sources like roads, parking lots, wastewater, and sewage treatment plants. For decades, several risky waste sites

M. P. Ajith · R. Paulraj (✉)
School of Environmental Sciences, Jawaharlal Nehru University, New Delhi 110067, India
e-mail: paulrajr@mail.jnu.ac.in

J. K. Katiyar et al. (eds.), *Nanomaterials for Advanced Technologies*,
https://doi.org/10.1007/978-981-19-1384-6_7

and industrial facilities have been polluted, causing environmental damage. Dams and regulator cum flyover have stopped the usual current of inland bodies of water, causing them to become standing. This has had several negative consequences in water bodies, including nutrient build-up and biomagnification, algal blooms, aquatic creature mortality, and salt intrusion (Ajith and James 2016).

Bioremediation is a popular and effective cleaning technology for removing hazardous waste from contaminated environments. Bioremediation is the process of degrading, eradicating, immobilizing, or detoxifying various chemical wastes and harmful physical elements from the environment using all-inclusive and actions of microorganisms. The conventional bioremediation strategies are not efficient enough to meet the real-world environmental scenario. These strategies showed a scope of improvement, which is now being addressed by the emerging fields of nanotechnology, which are efficient, cost-effective, and relatively less toxic. "Due to the complexity of the mixing of various chemicals, high volatility, and low reactivity of environmental contaminants, recent research has focused on the use of nanomaterials for the development of novel environmental remediation methods" (Tratnyek and Johnson 2006).

Nanomaterials have tremendously more surface-to-volume ratio than their bulkier equivalents, resulting in increased reactivity and greater efficacy. In addition, compared to traditional methods, nanomaterials can exploit unique surface chemistry, allowing them to be that can target specific contaminants for effective clean-up. The periodic adjustment of the physical parameters of nanoparticles, like size, shape, porosity, and chemical structure, might bestow other beneficial qualities that directly impact the material's effectiveness for contamination remediation. Approaches that combine many distinct materials (hybrids/composites) to extract desirable qualities from each component may be more effective, selective, and stable than methods relying on a single nano-platform. Material that has been functionalized with particular chemicals that target pollutants of interest can assist and improve the material's selectivity and efficiency (Guerra et al. 2017; Shah and Imae 2016; Campbell et al. 2015). This chapter discusses the most recent advancements of functionalized nanomaterials and nanocomposites that are used to decontaminate a selected range of contaminants in the environment.

2 Nanoparticles Used in Decontamination Technology

The tremendous advancement in nanotechnology helped us fabricate the nanoparticles in the desired properties that we are looking for. This made nanoparticles a commercially important material. In manufacturing industries, nanoparticles have been widely utilized because of their exclusive wide array of properties like electronic properties, catalytic activity, surface functionality and charge, magnetic properties, optical properties, antimicrobial activities, etc. The recent shift to biogenic synthesis of nanoparticles made the nanoparticle environmentally safe and economical.

Recently, nanotechnology has been extensively employed for the decontamination of water, soil, and air. Nanotechnology even made it possible to integrate our conventional remediation strategies with the nanobiotechnological approach. This enables us to bring a better efficacy in the removal of pollutants. The advanced research shows nanotechnology's promising potential either directly or indirectly integrating nanotechnology with other fields.

There is a variety of nanoparticles used in remediation. Its applications include decontamination of heavy metals, pesticides, endocrine disruptors, dyes, other hydrocarbons, solid wastes, and radioactive material. The nanoparticles, their associated devices, and technologies are employed for specific decontamination applications dependent on the particular properties. The nanoparticles used for the water remediation can be broadly classified into organic nanoparticles, inorganic nanoparticles, and polymer-based nanoparticles (Ajith et al. 2021).

3 Organic Nanoparticles

There are unique electronic, chemical, and physical properties for organic compounds because of elemental carbon's structural and tunable hybridization state. The primary nanostructural forms of carbon include nanostructures single-walled and multi-walled nanotubes (SWNT and MWNT), graphene, and fullerene C_{60} and C_{540} (Mauter and Elimelech 2008).

The surface functionalization and activation of carbon materials are the prior requirements for employing them in water remediation. SWNT, MWNT, and graphene are widely used for these purposes. The exclusive adsorption properties of these materials make it as an excellent material existing for adsorption of organic contaminants and inorganic contaminants in aqueous solutions (Ren et al. 2011; Theron et al. 2008; Kharisov et al. 2014). Both chemisorption and physisorption are observed based on the functionalization and the fabrication method. Chemisorption is a type of adsorption in which the surface and the adsorbate undergo a chemical interaction. At the adsorbent surface, new chemical bonds are formed.

In contrast to chemisorption, physisorption preserves the chemical species of the adsorbate and the surface (Fig. 2). Typically, chemisorption results in monolayer adsorption, whereas physisorption results in a multilayer arrangement of substrate

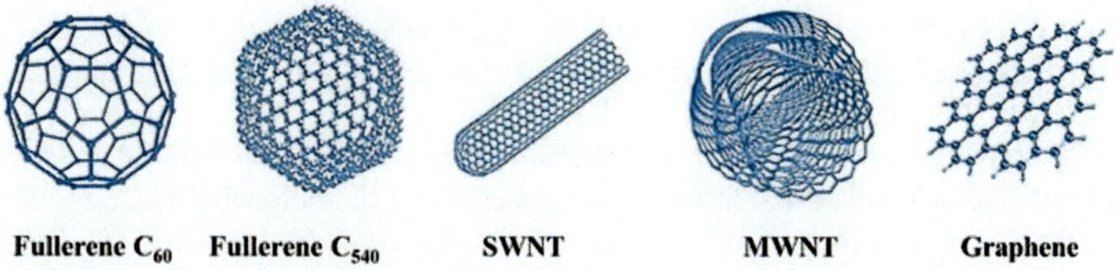

Fig. 1 Primary nanostructural forms of carbon

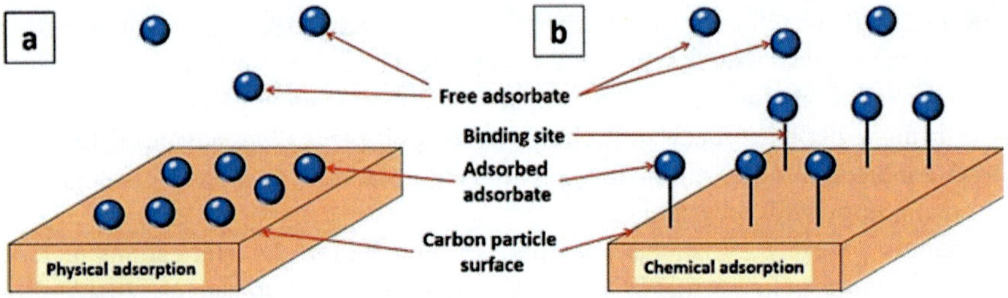

Fig. 2 Illustration of the mechanism of carbon particle surface–adsorbent interaction in physisorption and chemisorption. Reproduced with permission from Nandiyanto et al. (2020)

molecules (Nandiyanto et al. 2020). A previous study from our laboratory showed the efficiency of carbon quantum dots (CQD) in the remediation of heavy metals from industrial run-offs by engineering a silica–CQD nano-conjugate packed short-bed adsorber column. We reported a double efficiency with the conjugate than with the silica alone. Carbon-based nanoparticles are now extensively utilized in the decontamination of electroplating effluent, dye industries run-offs, wastewater remediation, sewage treatment, etc. (MP et al. 2022). The non-toxicity of these nanomaterials is an extra benefit for their use in clean-up. There is no biomagnification and toxicity-induced physiological abnormalities in trophic levels (Ajith et al. 2020).

Apart from the adsorption, one of the successfully used mechanisms for remediation is photocatalysis. Irradiation of UV on graphene, SWNT, and MWNT with photon energy more than or equal to the bandgap will enhance the formation of conductance band electron and valance band hole. The hole thereby generates hydroxyl radical, which helps oxidize chlorine-containing organic compounds (Gangu et al. 2019). The electron can facilitate the reduction of organic contaminants (Fig. 3). The photocatalyst using graphene nanocomposite (Yang et al. 2013; Chowdhury and Balasubramanian 2014; Liu et al. 2011; Zhang et al. 2010) also trends to enhance the adsorption efficiency further. Titanium oxide nanoparticle coupled with graphene has shown an enhancement in photocatalyst compared with titanium oxide nanoparticle alone (Zhang et al. 2010).

3.1 *SWNTs and MWNTs*

The SWNTs are mostly hexagonal with heterogeneous porous structures. There are two types of possible absorption in general open-ended SWNTs. "The materials with lower adsorption efficiency will bind on the external surface of the bundle of SWNTS, where those with higher adsorption efficacy can adsorb intra-tubularly or inter-tubularly. The external surface adsorption reaches saturation and hence equilibrium speedier than the intra-tubular and inter-tubular sites" (Ren et al. 2011). In the case of MWNTs, they do not exist as bundles unless they are prepared in exceptional methods.

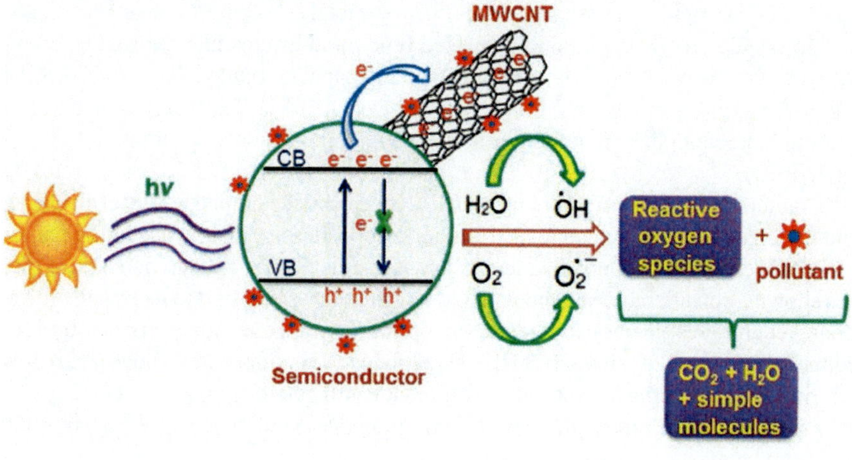

Fig. 3 Schematics of photocatalytic degradation by MTCNTs. Reproduced with permission from Gangu et al. (2019)

Another critical factor that affects the adsorption of CNTs is oxygen availability. The surface functional groups such –OH, –COOH, and –C=O can enhance the adsorption. To oxidize CNTs, wide varieties of compounds such as HNO_3, H_2SO_4, $KMnO_4$, NaOCl, and H_2O_2 can be utilized. There are reports of enhanced adsorption of oxidized CNTs with heavy metals (Khin et al. 2012). "The molecular weight of adsorbate, dipole movement, critical temperature, and pH is also a significant reason which affects the adsorption of CNTs. These factors affect the electrostatic attraction between the CNTs surface and the positively charged adsorbates such as heavy metals, dyes, and other organic compounds" (Lithoxoos et al. 2010). The efficiency of MWNTS in bacterial adsorption is reported by Lemes et al. (2011).

Ncibi and Sillanpää (2015) investigated the efficiency of single-walled, double-walled, and multi-walled carbon nanotube (SWCNT, DWCNT, MWCNT), agglomerate for the remediation of antibiotic ciprofloxacin, and oxytetracycline from aqueous medium. They found that SWCNT has the highest removal efficiency for both antibiotics.

3.2 Graphene

"Graphene, surface modified graphene, and graphene conjugates are used in water remediation. The surface modification of graphene prevents the aggregation of graphene layers and provides more surface area for adsorption" (An and Jimmy 2011). Because of this, surface-functionalized graphene is more capable than pure graphene. Graphene oxide is a modified graphene that is extensively utilized for the adsorption of an extensive choice of organic and inorganic aqueous impurities likes medicines, pesticides, pharmaceuticals, heavy metals, volatile organic chemicals, and so on. The oxygen-containing surface functional groups like hydroxyls, epoxides, carboxylic acids, and others contribute to graphene oxide's higher absorption efficiency as compared to pure graphene (Wang et al. 2013). These functional groups can facilitate acid–base reaction with basic contaminants such as ammonia (Wang et al. 2013).

Titanium oxide graphene (Zhang et al. 2010), zinc oxide graphene (Liu et al. 2011), and CdS–graphene (Zhang et al. 2013) nanocomposites are mainly used with photocatalytic activities to remove aqueous contaminants. The graphene nanocomposite has shown a considerable enhancement in its efficiency compared to its pristine form. Zhang et al. (2010) demonstrated that an optimum graphene balance is required for better catalytic activity in titanium oxide graphene nanocomposite. Above and below this optimum ratio, the photocatalytic efficiency will be reduced. Zhang et al. (2013) also shown a coinciding outcome of CdS–graphene nanocomposite by stating the optimum balance of graphene in nanocomposite for enhanced photocatalytic degradation. Liu et al. (2011) demonstrated a 40% enhancement in reducing chromium (VI) by zinc oxide graphene nanocomposite compared to the pure zinc oxide.

4 Inorganic Nanoparticles

4.1 Metal and Metal Oxide Nanoparticle

A variety of metal and metal oxide NPs (MONPs) are utilized in the decontamination. It is highly efficient in adsorbing heavy metals and chlorinated organic water contaminants. Due to their fast kinetics, adsorption capacity, and adaptability in both in situ and ex situ applications, MNOPs are an unavoidable component in water remediation, particularly in aquatic systems (Das et al. 2015; Santhosh et al. 2016).

Titanium oxide nanoparticles (TiO_2 NPs) are a promising metal oxide nanoparticle employed in water remediation. "The exceptional characteristics such as non-toxicity, low cost, semiconducting property, energy converting properties, and photocatalytic activity facilitate it to widely use in water treatment, surface-cleaning, air purification, waste treatment, etc." (Li et al. 2008; Adesina 2004). Photoactivation of TiO_2 NPs generates reactive oxidants like hydroxyl radical, which serve as a disinfectant for microbes (Zan et al. 2007; Cho et al. 2005). Guan et al. (2012) summarized TiO_2-based arsenic removal approaches based on photocatalytic oxidation of arsenite/organic arsenic to arsenate and inorganic and organic arsenic adsorption. Figure 4 shows the representation of TiO_2 NPs application in arsenic elimination. Many studies have reported that doping TiO_2 NPs with other transmission elements enhances their photocatalytic activity. For example, Ag-doped TiO_2 nanofibers have shown a noteworthy enhancement in photocatalytic breakdown of 2-chlorophenol (Park and Lee 2014). Studies are also offering the TiO_2 NPs mixed with SiO_2 as an effective photocatalytic agent for breakdown of dyes (Rasalingam et al. 2014).

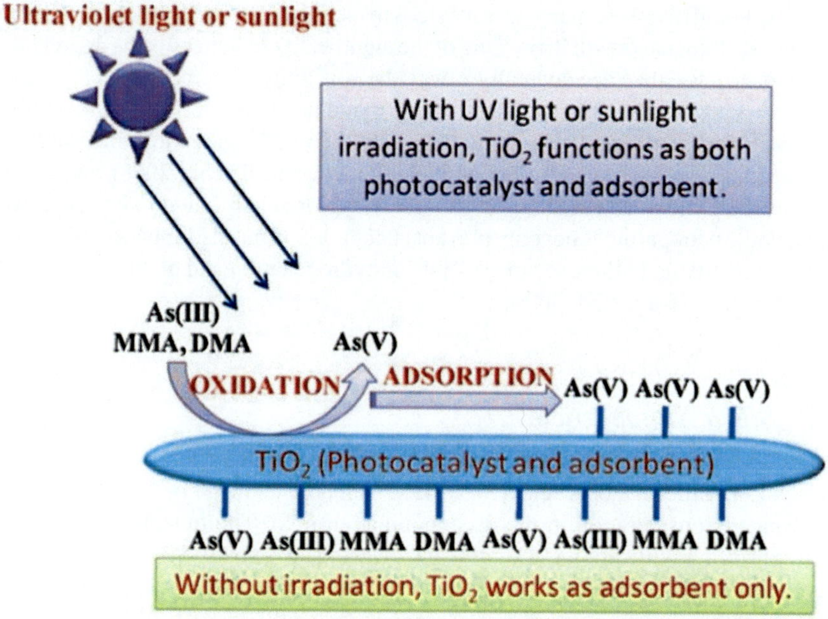

Fig. 4 Schematics of TiO_2 NPs in arsenic removal. Reproduced with permission from Guan et al. (2012)

Silver nanoparticles (AgNPs) are another crucial metal nanoparticle widely employed in water disinfection (Gupta and Silver 1998). The antibacterial, antifungal, and antiviral activity of AgNPs made it a significant one in water treatment. It is found that the AgNPs below 10 nm is more effective than the larger particles. They prevent the binding of the virus to the host cells by binding to the glycoproteins of the virus. They are also toxic to pathogenic bacteria like *Pseudomonas aeruginosa* and *Escherichia coli.* The shape and surface topography of the AgNPs also affect adsorption behaviour (Pal et al. 2007).

Iron and iron oxide nanoparticles are extensively used nanoparticles because of their efficiency in removal of Cd^{2+} (Ebrahim et al. 2016), Cu^{2+} (Poguberović et al. 2016), Co^{2+} (Hooshyar et al. 2013), and Ni^{2+} (Hooshyar et al. 2013) and chlorinated organic pollutants (Guo et al. 2017). Iron NPs exist as a core–shell structure. Core consists of zerovalent Fe, and shell consists of mixed Fe (II) and Fe (III) oxides (Kharisov et al. 2012). Remediation is facilitated by electron donation of the core zerovalent Fe, thereby reducing the heavy metals and chlorinated organic contaminants. The shell enhances the remediation by reducing the pollutants present in the higher standard reduction potential (Li and Zhang 2006). The iron oxide nanoparticle can easily retain and be separated from the system after the treatment because of their magnetic property. This is an added advantage, which makes metallic nanoparticle more promising one in water remediation.

Even though there are many positive sides to metallic nanoparticles, there are some challenges associated with them. One of the significant challenges is the aggregation of nanoparticles that detrimentally affect the reactivity. Another one is the toxicity associated with them. Metal and metal oxide nanoparticles can even enter the food web and can cause biomagnification. To prevent aggregation, incorporating second metal such as nickel (Tee et al. 2005; Wu and Ritchie 2006) and copper (Zheng et al. 2009; Zhu et al. 2010) is widely explored and found to be successful. "Bimetallic nanoparticles not only prevent accumulation but also enhance the zerovalent Fe stability and efficiency in catalysis of hydrogenation and dichlorination with contaminants" (Karn et al. 2009).

4.2 Silica Nanoparticle

Mesoporous silica nanoparticles (SiO_2 NPs) with pore diameter of 2–50 nm grabbed widespread consideration in the decontamination field. The tunable pore size and facile surface modification is a significant advantage to employ SiO_2 NPs in a diversity of fields. The surface hydroxyl groups present in SiO_2 NPs facilitate the surface modifications.

Silica nanoparticles are efficient in photodegradation dyes like methyl red azo dye (Das et al. 2013). According to these research, SiO_2 NPs have more effective photocatalytic property than SiO_2 NPs coated with Ag NPs and Au NPs. SiO_2 nanoparticles improve oil recovery and decrease the discharge of heavy metals, radioactive chemicals, and brine into water. The functionality of SiO_2 NPs must be altered in order to balance and support the oil extraction process. SiO_2 NPs were treated with H+ to create a shielding effect in saltwater for oil extraction (Shi et al. 2013).

5 Polymer-Based Nanomaterials

Although NPs offer several advantages, as previously stated, one of their restrictions is their stability. Many reports signpost that following fabrication, NPs can aggregate based on a diversity of circumstances. "The aggregation of particles detrimentally affects the adsorption ability. An alternative for this is the use of polymer-based nanomaterials. Here we use the matrix or support material to hold the nanomaterials, which prevents aggregation and enhances the stability of pristine nanoparticles" (Ajith et al. 2021). The polymeric host includes stabilizing agents, surface-functionalized ligands, emulsifiers, and surfactants. Polymer-based nanoparticles are used in sensing as well as the decontamination of a diversity of pollutants such as heavy metals (e.g. Fe, Mn, As, etc.), organic chemicals (e.g. pesticides, pharmaceuticals, volatile organic compounds, other aliphatic and aromatic compounds), gases (e.g. SO_2, CO_2, NO_2), and microbes (e.g. bacteria, viruses, and other pathogens).

The nanosponges formed from cyclodextrin-based highly cross-linked polymer are effective in removing 3,5,6-trichloro-2-pyridinyloxyacetic acid (triclopyr) from soil with a removal efficiency of 92% (Baglieri et al. 2013). Das et al. (2017) fabricated polypyrrole (PPy) conjugated with mercaptoacetic acid (MAA) and shown its effectiveness in the remediation of aluminium (Ag^+) from contaminated water. They noticed more efficiency in adsorption for composite than the pristine PPy.

Dendrimers are polymers made up of monodispersed, highly branching macromolecules. Dendritic NPs are extensively utilized in the treatment of heavy metal-polluted wastewater. Dendritic nanoparticles with functional groups like carboxylates, hydroxyl groups, hydroxamates, and amines aid in the encapsulation of a verity of contaminants, particularly cations such as Zn^{2+}, Cu^{2+}, Au^{2+}, Ag^{2+}, Mg^{2+}, Mn^{2+}, Fe^{2+}, Fe^{3+}, Ni^{2+}, and others. (Diallo et al. 2005). Compared to the linear polymers, the dendritic nanolayers do not pass easily via the ultra-filtration membrane, which is an advantage for water decontamination.

Among the several concepts outlined, tightly packed NPs and nanofiber membranes, arranged nanotube membranes, self-assembled 2D layer nanomaterials, and their composites are among the most capable membrane technologies (Ying et al. 2016; Huang et al. 2015; Liu et al. 2016). Figure 5 depicts the most typical nanomaterial-based membrane architectures (Ying et al. 2017).

Polymer-based nanomaterials are also found to be a potent antimicrobial agent. Qasim et al. (2018) described the antimicrobial property of $AgNO_3$ NPs encapsulated poly-isopropyl acrylamide-based polymeric nanoparticles. The polymer is

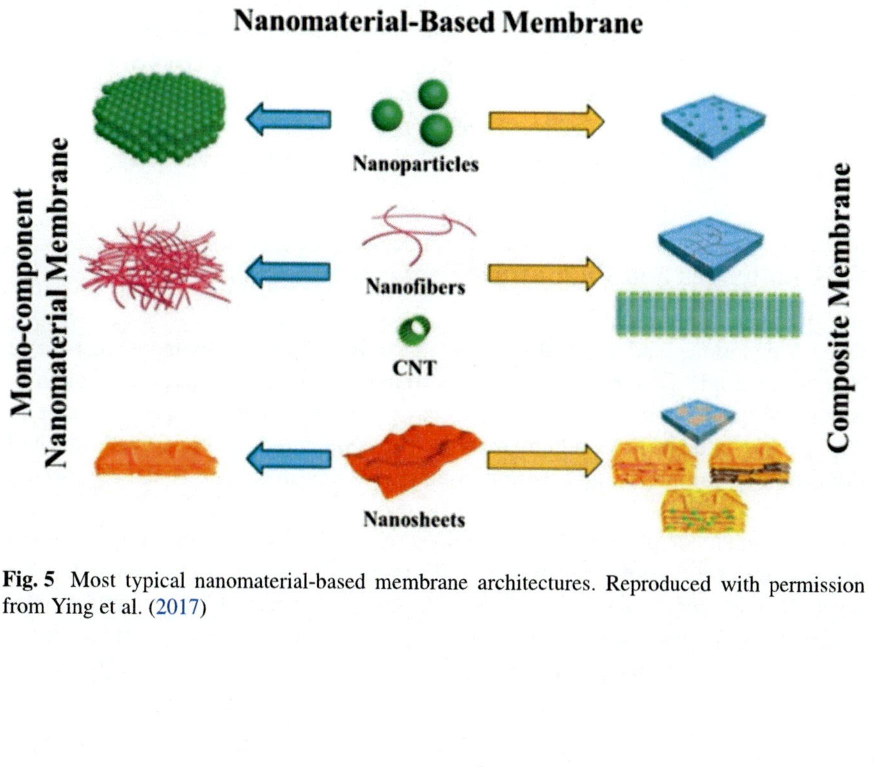

Fig. 5 Most typical nanomaterial-based membrane architectures. Reproduced with permission from Ying et al. (2017)

effective in immobilizing both gram-positive and gram-negative bacteria dependent on the amount and size of $AgNO_3$ used during encapsulation. An earlier study by Dhanaseker et al. (2018) showed that photodegradation of microorganisms using Cu-doped TiO_2 NPs (Cu_2O-TiO_2/rGO) with solid support graphene oxide. They concluded that Cu_2O-TiO_2/rGO displays an exceptional visible light photocatalytic antimicrobial activity with low minimum inhibitory concentration (MIC) for gram-positive and gram-negative bacteria.

Metal and metal oxide-based polymer nanocomposites were reported in previous studies showing their high reactivity (Ng et al. 2013). The significant metals used for metal-based polymer nanocomposites in water remediation include titanium, iron, silver, magnesium, copper, etc. Polyamide nanofibers modified with Al_2O_3, TiO_2, and other oxides are used in the air and water treatment to remove an array of pollutants (Bianco et al. 2007). These types of electrospun nanofibers are extensively employed for the remediation of toxic air. MgO nanoparticles conjugated with nanofiber membranes such as polysulfone (PSU), polyvinyl chloride (PVC), and polyvinylidene fluoride-co-hexafluoropropylene are used in the removal of toxic gas (Homaeigohar and Elbahri 2014; Sundarrajan and Ramakrishna 2007). Bimetallic systems are also found to be effective in water remediation. Kumar et al. (2020) reported that carboxymethyl cellulose (CMC) stabilized with iron and palladium is more effective in the removal of hexachlorocyclohexane (HCH) from the polluted soil than the pristine CMC. Here these metals act as a catalyst in the adsorption process of the system.

Magnetic polymer nanocomposites outperform metallic and metal oxide-based polymer nanocomposites in terms of biocompatibility and ease of removal from the treatment system via magnetic fields. The high surface energy of these nanocomposites can easily make them oxidized. To prevent that, the coating is done with silica, carbon, or some other polymers. The magnetic polymer nanocomposites are used for remediation of heavy metals such as Hg, Co, Cu, and Cr and organic dyes like methyl orange and methylene blue (Zhao et al. 2011; Rocher et al. 2008).

6 Challenges and the Future Concern

Due to economic and methodological constraints, high deployment costs, and possible ecological and human hazards, current nanotechnological research of environmental decontamination is limited to laboratory and model studies (Khan et al. 2019). Many studies have documented the widespread usage of nanomaterials and their toxicity to humans and the environment. A key drawback is the absence of mechanistic examination of nanomaterials used in environmental pollution. For a long time, nanocomposites of functionalized nanomaterials have been deposited in the environment, resulting in biomagnification and entry into the higher trophic levels of ecosystems (Gehrke et al. 2015). Along with advanced research in this area, research should also focus on strategies to prevent nanomaterial deposition in the environment and manage or balance ecosystems without causing harmful effects.

The fabrication of biogenic nanomaterials with characteristics such as long-term viability, cheap cost, and ecological sustainability might play a substantial part in future remediation of wastewaters.

7 Conclusion

The present chapter delivers a thorough impression of the existing state-of-the-art nanomaterials utilized for sustainable environmental decontamination, concentrating on their applications in innovative remedial methods such as adsorption/filtration, catalysis, photodegradation, and electro-nano-remediation. Insights on the performance of processes, the modes of deployment, potential risk and their management, and the consequences of nanomaterial utilization for environmental decontamination for society and the economy indicate that the broad acceptance of nano-remediation technologies requires not only substantial advancements of the underlying aspects of science and engineering but also an actual display of efficiency in the real world using already established techniques.

Acknowledgements The Department of Science and Technology, New Delhi, India, provided funding to AMP through the INSPIRE Senior Research Fellowship which is gratefully acknowledged.

References

A.A. Adesina, Industrial exploitation of photocatalysis: progress, perspectives and prospects. Catal. Surv. Asia **8**(4), 265–273 (2004)

M.P. Ajith, M.K. James, Impact of Chamravattam regulator cum bridge on Bharathapuzha river and adjacent areas. Indian J. Econ. Dev. **4**(1), 2320–9828 (2016)

M.P. Ajith, M. Aswathi, E. Priyadarshini, P. Rajamani, Recent innovations of nanotechnology in water treatment: a comprehensive review. Bioresour. Technol. 126000 (2021)

M.P. Ajith, E. Priyadarshini, P. Rajamani, Effective and selective removal of heavy metals from industrial effluents using sustainable Si–CD conjugate based column chromatography. Bioresour. Technol. **314**, 123786 (2020)

X. An, C.Y. Jimmy, Graphene-based photocatalytic composites. Rsc Adv. **1**(8), 1426–1434 (2011)

A. Baglieri, M. Nègre, F. Trotta, P. Bracco, M. Gennari, Organo-clays and nanosponges for acquifer bioremediation: adsorption and degradation of triclopyr. J. Environ. Sci. Health B **48**(9), 784–792 (2013)

A. Bianco, G. Iardino, C. Bertarelli, L. Miozzo, A. Papagni, G. Zerbi, Modification of surface properties of electrospun polyamide nanofibers by means of a perfluorinated acridine. Appl. Surf. Sci. **253**(20), 8360–8364 (2007)

M.L. Campbell, F.D. Guerra, J. Dhulekar, F. Alexis, D.C. Whitehead, Target-specific capture of environmentally relevant gaseous aldehydes and carboxylic acids with functional nanoparticles. Chem. Euro. J. **21**(42), 14834–14842 (2015)

M. Cho, H. Chung, W. Choi, J. Yoon, Different inactivation behaviors of MS-2 phage and *Escherichia coli* in TiO_2 photocatalytic disinfection. Appl. Environ. Microbiol. **71**(1), 270–275 (2005)

S. Chowdhury, R. Balasubramanian, Recent advances in the use of graphene-family nanoadsorbents for removal of toxic pollutants from wastewater. Adv. Coll. Interface. Sci. **204**, 35–56 (2014)

R. Das, S. Giri, A.L. King Abia, B. Dhonge, A. Maity, Removal of noble metal ions (Ag+) by mercapto group-containing polypyrrole matrix and reusability of its waste material in environmental applications. ACS Sustain. Chem. Eng. **5**(3), 2711–2724 (2017)

S. Das, B. Sen, N. Debnath, Recent trends in nanomaterials applications in environmental monitoring and remediation. Environ. Sci. Pollut. Res. **22**(23), 18333–18344 (2015)

S.K. Das, M.M.R. Khan, T. Parandhaman, F. Laffir, A.K. Guha, G. Sekaran, A.B. Mandal, Nano-silica fabricated with silver nanoparticles: antifouling adsorbent for efficient dye removal, effective water disinfection and biofouling control. Nanoscale **5**(12), 5549–5560 (2013)

M. Dhanasekar, V. Jenefer, R.B. Nambiar, S.G. Babu, S.P. Selvam, B. Neppolian, S.V. Bhat, Ambient light antimicrobial activity of reduced graphene oxide supported metal doped TiO_2 nanoparticles and their PVA based polymer nanocomposite films. Mater. Res. Bull. **97**, 238–243 (2018)

M.S. Diallo, S. Christie, P. Swaminathan, J.H. Johnson, W.A. Goddard, Dendrimer enhanced ultrafiltration. 1. Recovery of Cu (II) from aqueous solutions using PAMAM dendrimers with ethylene diamine core and terminal NH_2 groups. Environ. Sci. Technol. **39**(5), 1366–1377 (2005)

S.E. Ebrahim, A.H. Sulaymon, H. Saad Alhares, Competitive removal of Cu^{2+}, Cd^{2+}, Zn^{2+}, and Ni^{2+} ions onto iron oxide nanoparticles from wastewater. Desalin. Water Treat. **57**(44), 20915–20929 (2016)

K.K. Gangu, S. Maddila, S.B. Jonnalagadda, A review on novel composites of MWCNTs mediated semiconducting materials as photocatalysts in water treatment. Sci. Total Environ. **646**, 1398–1412 (2019)

I. Gehrke, A. Geiser, A. Somborn-Schulz, Innovations in nanotechnology for water treatment. Nanotechnol. Sci. Appl. **8**, 1 (2015)

X. Guan, J. Du, X. Meng, Y. Sun, B. Sun, Q. Hu, Application of titanium dioxide in arsenic removal from water: a review. J. Hazard. Mater. **215**, 1–16 (2012)

F.D. Guerra, M.L. Campbell, D.C. Whitehead, F. Alexis, Tunable properties of functional nanoparticles for efficient capture of VOCs. ChemistrySelect **2**(31), 9889–9894 (2017)

M. Guo, X. Weng, T. Wang, Z. Chen, Biosynthesized iron-based nanoparticles used as a heterogeneous catalyst for the removal of 2, 4-dichlorophenol. Sep. Purif. Technol. **175**, 222–228 (2017)

A. Gupta, S. Silver, Silver as a biocide: will resistance become a problem? Nat. Biotechnol. **16**(10), 888–888 (1998)

S. Homaeigohar, M. Elbahri, Nanocomposite electrospun nanofiber membranes for environmental remediation. Materials **7**(2), 1017–1045 (2014)

Z. Hooshyar, G. RezanejadeBardajee, Y. Ghayeb, Sonication enhanced removal of nickel and cobalt ions from polluted water using an iron-based sorbent. J. Chem. (2013)

L. Huang, M. Zhang, C. Li, G. Shi, Graphene-based membranes for molecular separation. J. Phys. Chem. Lett. **6**(14), 2806–2815 (2015)

B. Karn, T. Kuiken, M. Otto, Nanotechnology and in situ remediation: a review of the benefits and potential risks. Environ. Health Perspect. **117**(12), 1813–1831 (2009)

N.A. Khan, S.U. Khan, S. Ahmed, I.H. Farooqi, A. Dhingra, A. Hussain, F. Changani, Applications of nanotechnology in water and wastewater treatment: a review. Asian J. Water Environ. Pollut. **16**(4), 81–86 (2019)

B.I. Kharisov, H.R. Dias, O.V. Kharissova, Nanotechnology-based remediation of petroleum impurities from water. J. Petrol. Sci. Eng. **122**, 705–718 (2014)

B.I. Kharisov, H.R. Dias, O.V. Kharissova, V.M. Jiménez-Pérez, B.O. Perez, B.M. Flores, Iron-containing nanomaterials: synthesis, properties, and environmental applications. RSC Adv. **2**(25), 9325–9358 (2012)

M.M. Khin, A.S. Nair, V.J. Babu, R. Murugan, S. Ramakrishna, A review on nanomaterials for environmental remediation. Energy Environ. Sci. **5**(8), 8075–8109 (2012)

N. Kumar, A. Balamurugan, P. Balakrishnan, K. Vishwakarma, K. Shanmugam, biogenic nanomaterials: synthesis and its applications for sustainable development. Biogenic Nano-Particles and their Use in Agro-ecosystems, 99–132 (2020). https://doi.org/10.1007/978-981-15-2985-6_7

A.P. Lemes, L. Cordi, A. Santos, N. Durán, Bacterial remediation from effluent containing multi-walled carbon nanotubes. J. Phys. Conf. Ser. **304**(1), 012023 (2011)

Q. Li, S. Mahendra, D.Y. Lyon, L. Brunet, M.V. Liga, D. Li, P.J. Alvarez, Antimicrobial nanomaterials for water disinfection and microbial control: potential applications and implications. Water Res. **42**(18), 4591–4602 (2008)

X.Q. Li, W.X. Zhang, Iron nanoparticles: the core−shell structure and unique properties for Ni (II) sequestration. Langmuir **22**(10), 4638–4642 (2006)

G.P. Lithoxoos, A. Labropoulos, L.D. Peristeras, N. Kanellopoulos, J. Samios, I.G. Economou, Adsorption of N_2, CH_4, CO and CO_2 gases in single walled carbon nanotubes: a combined experimental and Monte Carlo molecular simulation study. J. Supercrit. Fluids **55**(2), 510–523 (2010)

G. Liu, W. Jin, N. Xu, Two-dimensional-material membranes: a new family of high-performance separation membranes. AngewandteChemie Int. Ed. **55**(43), 13384–13397 (2016)

X. Liu, L. Pan, T. Lv, T. Lu, G. Zhu, Z. Sun, C. Sun, Microwave-assisted synthesis of ZnO–graphene composite for photocatalytic reduction of Cr (VI). Catal. Sci. Technol. **1**(7), 1189–1193 (2011)

M.S. Mauter, M. Elimelech, Environmental applications of carbon-based nanomaterials. Environ. Sci. Technol. **42**(16), 5843–5859 (2008)

A. MP, S. Pardhiya, P. Rajamani, Carbon dots: an excellent fluorescent probe for contaminant sensing and remediation. Small 2105579 (2022).https://doi.org/10.1002/SMLL.202105579

A.B.D. Nandiyanto, G.C.S. Girsang, R. Maryanti, R. Ragadhita, S. Anggraeni, F.M. Fauzi, P. Sakinah, A.P. Astuti, D. Usdiyana, M. Fiandini, M.W. Dewi, Isotherm adsorption characteristics of carbon microparticles prepared from pineapple peel waste. Commun. Sci. Technol. **5**(1), 31–39 (2020)

M.C. Ncibi, M. Sillanpää, Optimized removal of antibiotic drugs from aqueous solutions using single, double and multi-walled carbon nanotubes. J. Hazard. Mater. **298**, 102–110 (2015). https://doi.org/10.1016/J.JHAZMAT.2015.05.025

L.Y. Ng, A.W. Mohammad, C.P. Leo, N. Hilal, Polymeric membranes incorporated with metal/metal oxide nanoparticles: a comprehensive review. Desalination **308**, 15–33 (2013)

S. Pal, Y.K. Tak, J.M. Song, Does the antibacterial activity of silver nanoparticles depend on the shape of the nanoparticle? A study of the gram-negative bacterium *Escherichia coli*. Appl. Environ. Microbiol. **73**(6), 1712–1720 (2007)

J.Y. Park, I.H. Lee, Photocatalytic degradation of 2-chlorophenol using Ag-doped TiO_2 nanofibers and a near-UV light-emitting diode system. J. Nanomater. (2014)

S.S. Poguberović, D.M. Krčmar, S.P. Maletić, Z. Kónya, D.D.T. Pilipović, D.V. Kerkez, S.D. Rončević, Removal of As (III) and Cr (VI) from aqueous solutions using "green" zero-valent iron nanoparticles produced by oak, mulberry and cherry leaf extracts. Ecol. Eng. **90**, 42–49 (2016)

M. Qasim, N. Udomluck, J. Chang, H. Park, K. Kim, Antimicrobial activity of silver nanoparticles encapsulated in poly-N-isopropylacrylamide-based polymeric nanoparticles. Int. J. Nanomed. **13**, 235–249 (2018). https://doi.org/10.2147/IJN.S153485

S. Rasalingam, R. Peng, R.T. Koodali, Removal of hazardous pollutants from wastewaters: applications of TiO_2-SiO_2 mixed oxide materials. J. Nanomater. (2014)

X. Ren, C. Chen, M. Nagatsu, X. Wang, Carbon nanotubes as adsorbents in environmental pollution management: a review. Chem. Eng. J. **170**(2–3), 395–410 (2011)

V. Rocher, J.M. Siaugue, V. Cabuil, A. Bee, Removal of organic dyes by magnetic alginate beads. Water Res. **42**(4–5), 1290–1298 (2008)

C. Santhosh, V. Velmurugan, G. Jacob, S.K. Jeong, A.N. Grace, A. Bhatnagar, Role of nanomaterials in water treatment applications: a review. Chem. Eng. J. **306**, 1116–1137 (2016)

K.J. Shah, T. Imae, Selective gas capture ability of gas-adsorbent-incorporated cellulose nanofiber films. Biomacromolecules **17**(5), 1653–1661 (2016)

S. Shi, W. Wang, L. Liu, S. Wu, Y. Wei, W. Li, Effect of chitosan/nano-silica coating on the physicochemical characteristics of longan fruit under ambient temperature. J. Food Eng. **118**(1), 125–131 (2013)

S. Sundarrajan, S. Ramakrishna, Fabrication of nanocomposite membranes from nanofibers and nanoparticles for protection against chemical warfare stimulants. J. Mater. Sci. **42**(20), 8400–8407 (2007)

Y.H. Tee, E. Grulke, D. Bhattacharyya, Role of Ni/Fe nanoparticle composition on the degradation of trichloroethylene from water. Ind. Eng. Chem. Res. **44**(18), 7062–7070 (2005)

J. Theron, J.A. Walker, T.E. Cloete, Nanotechnology and water treatment: applications and emerging opportunities. Crit. Rev. Microbiol. **34**(1), 43–69 (2008)

P.G. Tratnyek, R.L. Johnson, Nanotechnologies for environmental cleanup. Nano Today **1**(2), 44–48 (2006)

S. Wang, H. Sun, H.M. Ang, M.O. Tadé, Adsorptive remediation of environmental pollutants using novel graphene-based nanomaterials. Chem. Eng. J. **226**, 336–347 (2013)

L. Wu, S.M. Ritchie, Removal of trichloroethylene from water by cellulose acetate supported bimetallic Ni/Fe nanoparticles. Chemosphere **63**(2), 285–292 (2006)

M.Q. Yang, N. Zhang, Y.J. Xu, Synthesis of fullerene–, carbon nanotube–, and graphene–TiO_2 nanocomposite photocatalysts for selective oxidation: a comparative study. ACS Appl. Mater. Interfaces. **5**(3), 1156–1164 (2013)

Y. Ying, Y. Yang, W. Ying, X. Peng, Two-dimensional materials for novel liquid separation membranes. Nanotechnology **27**(33), 332001 (2016)

Y. Ying, W. Ying, Q. Li, D. Meng, G. Ren, R. Yan, X. Peng, Recent advances of nanomaterial-based membrane for water purification. Appl. Mater. Today **7**, 144–158 (2017)

L. Zan, W. Fa, T. Peng, Z.K. Gong, Photocatalysis effect of nanometer TiO_2 and TiO_2-coated ceramic plate on Hepatitis B virus. J. Photochem. Photobiol. B **86**(2), 165–169 (2007)

N. Zhang, M.Q. Yang, Z.R. Tang, Y.J. Xu, CdS–graphene nanocomposites as visible light photocatalyst for redox reactions in water: a green route for selective transformation and environmental remediation. J. Catal. **303**, 60–69 (2013)

Y. Zhang, Z.R. Tang, X. Fu, Y.J. Xu, TiO_2- graphene nanocomposites for gas-phase photocatalytic degradation of volatile aromatic pollutant: is TiO_2- graphene truly different from other TiO_2- carbon composite materials? ACS Nano **4**(12), 7303–7314 (2010)

X. Zhao, L. Lv, B. Pan, W. Zhang, S. Zhang, Q. Zhang, Polymer-supported nanocomposites for environmental application: a review. Chem. Eng. J. **170**(2–3), 381–394 (2011)

Z. Zheng, S. Yuan, Y. Liu, X. Lu, J. Wan, X. Wu, J. Chen, Reductive dechlorination of hexachlorobenzene by Cu/Fe bimetal in the presence of nonionic surfactant. J. Hazard. Mater. **170**(2–3), 895–901 (2009)

N. Zhu, H. Luan, S. Yuan, J. Chen, X. Wu, L. Wang, Effective dechlorination of HCB by nanoscale Cu/Fe particles. J. Hazard. Mater. **176**(1–3), 1101–1105 (2010)

Thermoelastic Vibrations of Functionally Graded Nonuniform Nanobeams

Rahul Saini

Abstract This chapter starts with the historical development of vibrations of structural elements, and then the discussions and classifications of micromechanical models for advanced composite materials, macro and micro/nano continuum mechanics theories and thermal field are presented. To study the dynamic behaviour of structural elements, a mathematical model is obtained for nonuniform functionally graded nanobeam subjected to the thermal environment using various physical assumptions. Further, numerical techniques are employed to obtain the approximate solution of this model. Eventually, a debate on quadrature methods to reveal their efficiency and convergence following the analysis of vibration characteristics of nonuniform functionally graded nanobeams under thermal field is presented.

Keywords Thermoelastic vibrations · Functionally graded models · Nonuniform nanobeams · Quadrature methods

1 Vibration

Vibration is a phenomenon of common occurrence in nature ranging from modern missile and space technology to household goods such as mixers, grinders and washing machines. Vibrations are considered desirable and undesirable depending upon their characteristics. Vibrations of teleprinters, telephone receivers, conveyors, compactors and strings in musical instruments are desirable, while that for wind turbine blades, helicopter rotors, aircraft wings and rudders, bridges and dams, high-rise buildings may be disastrous. Vibration may cause excessive noise due to the wear of machine parts leading to reduction in performance. Thus, the knowledge of the theory of vibration is essential for a design engineer not only to prevent undesirable vibrations and noise but also to increase the efficiency and life span of the machines. The vibration of elastic bodies has its beginning in the early thirties of the seventeenth century when Galileo observed that the frequency of a simple pendulum

R. Saini (✉)
Department of Applied Mechanics, Indian Institute of Technology Delhi, New Delhi 110016, India
e-mail: rahulsainiiit@gmail.com; rahul.saini@am.iitd.ac.in

J. K. Katiyar et al. (eds.), *Nanomaterials for Advanced Technologies*,
https://doi.org/10.1007/978-981-19-1384-6_8

depends upon its length and reported in "Discourses Concerning Two New Sciences". Later on, the correct account of the vibration of strings was presented in his book "Harmonicorum Liber" by Mersenne in 1636. Afterward, the experimental and theoretical developments on this issue have been carried out by many workers, and some of the pioneers are John Wallis, Robert Hook, Sir Isaac Newton, Sauveur, Brook Taylor, Daniel Bernoulli, Jean D'Alembert, Leonard Euler, Joseph Fourier, Joseph Lagrange, Leibnitz, Soedel (2004), Chladni (1787), Biot (1816), Bernoulli (1735), Euler (1744), Coulomb (1784), Cauchy (1827), Saint–Venant (1849), Poisson (1829) and Lamè (1852).

Owing to the various assumptions for the beams considered by the above researchers, various displacement theories such as Euler–Bernoulli beam theory (Nazemnezhad and Hosseini-Hashemi 2014; Farajpour et al. 2011; Ebrahimi et al. 2015; Nejad and Hadi 2016a, b; Nejad et al. 2016; Wang et al. 2016), Timoshenko beam theory (Şimşek 2015; Rahmani and Pedram 2014), higher-order shear deformation theory (Pradhan 2009), Reddy–Bickford beam theory, general exponential shear deformation beam theory and Levinson beam theory are proposed by various researchers (Wang et al. 2000). Among these, the Euler–Bernoulli beam theory is the simplest one and commonly used by engineers and scientists to predict the behaviour of beams. This theory is based upon some assumptions which are as follows:

- The thickness of the beam is small as compared to the length of the beam.
- The normal stresses in the transverse direction of the middle plane of the beam are taken to be negligibly small.
- The middle surface of the beam remains unstrained during deformation.
- The normal to the undeformed middle surface remains straight and normal to the deformed middle surface.

In this regard, the mathematical theory of elasticity by Love (1944) is one of the oldest and best books to develop the subject to the present level. After that, a number of books have appeared from time to time covering various aspects of bending and vibration of structural elements. Notable ones amongst them are Szilard (1974), Panc (1975), Timoshenko and Krieger (1984), Shames and Dym (1985), Yu (1996), Ventsel and Krauthammer (2001), Rao (2004), Kienzler et al. (2004), Soedal (2004), Reddy (2008), Petyt (2010) and Mao and Pietrzko (2013).

2 Functionally Graded Materials

Functionally graded materials (FGMs) are advanced engineering composites in which material properties vary from one surface to the other continuously. Generally, these materials are manufactured by the composition of two constituent materials to use as structural elements in high-temperature environment such as the aerospace industry (aircraft, rockets, missiles, etc.), nuclear reactors, piezoelectric and thermoelectric devices, and petroleum and petrochemical industries. The traditional FGMs are made with ceramic and metal constituents such that metal behaves as traditional composites at high temperature while ceramic has high thermal resistivity which improves the overall performance of the structure. A group of Japanese material scientists developed the FGMs in 1984 (Koizumi 1993) to use them as thermal barrier materials for aerospace structures. The fabricated material has the capacity of 2000 K on one surface and 1000 K on another surface with a thickness of 10 mm. Further, various projects and researches started to develop and use them in various technological situations at extremely high-temperature environment. In the meantime, the subject is developed according to the diverse and numerous potential applications of FGM structures which is available in open literature: metal–ceramic armour (Liu et al. 2003), thin-walled rotating blades (Librescu et al. 2005), sensors (Müller et al. 2003), actuators (Qiu et al. 2003), photodetectors (Paszkiewicz et al. 2007) and dental implant (Watari et al. 2004). The various aspects of FGMs are published in past decades and summarized by various researchers in their review articles (Tanigawa 1995; Markworth et al. 1995; Fuchiyama and Noda 1995; Noda 1999; Paulino et al. 2007). Being a material to work at extreme temperature, most of the early research studies of FGMs are focused on thermal analysis of static and dynamic problems of different structural elements. Recently, Swaminathan and Sangeetha (2017) presented an extensive review for thermal analysis of FGM plates up to 2016.

2.1 *Models of Functionally Graded Materials*

To find the effective mechanical properties of the functionally graded materials, the material scientists gave a variety of micromechanical models (Zuiker 1995; Reiter et al. 1997; Gasik 1998; Birman and Byrd 2007; Akbarzadeh et al. 2015; Kim and Paulino 2003) in terms of the mechanical properties of constituent materials. Generally, these homogenization models are based on the approximation of the shape and volume fraction of constituents in a domain. A comprehensive discussion of the homogenization models was presented by Shen (2016). Out of these, the power law function, Mori–Tanaka's scheme and exponential function are widely used by researchers. The mathematical expressions to compute the effective mechanical properties of the FGMs for these homogenization techniques are defined as follows:

(i) **Power law function** (Lal and Saini 2019a, b, c):

$$E_z = \{E^c - E^m\}V_c^p + E^m, \tag{1.1a}$$

$$\rho_z = \{\rho^c - \rho^m\}V_c^p + \rho^m, \tag{1.1b}$$

$$\alpha_z = \{\alpha^c - \alpha^m\}V_c^p + \alpha^m, \tag{1.1c}$$

$$k_z = \{k^c - k^m\}V_c^p + k^m, \tag{1.1d}$$

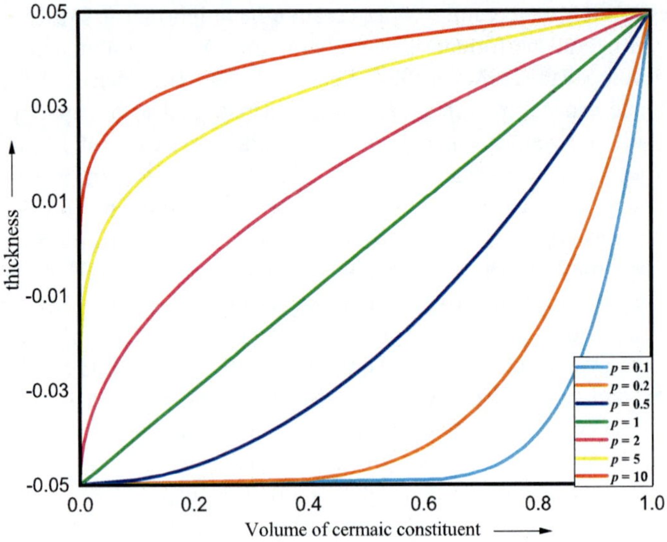

Fig. 1 Material distribution along the thickness of the beam

$$V_c = \frac{z}{h} + \frac{1}{2},$$

where superscripts c and m refer to ceramic and metal, respectively; E_z, ρ_z, α_z and k_z are the Young's modulus, mass density, thermal expansion coefficient and thermal conductivity, respectively, V_c is the volume of the ceramic constituent, and power p (≥ 0) controls the material distribution in the FG structure. The material distribution within the domain of the beam is shown in Fig. 1 for different values of p.

(ii) **Mori–Tanaka's scheme** (Sofiyev 2019; Fazzolari 2015):

$$K_z = K^m + \left(K^c - K^m\right)V_c^p\left[1 + \left(1 - V_c^p\right)\left(\frac{K^c - K^m}{K^m + \frac{4}{3}G^m}\right)\right]^{-1},$$

$$K^c = \frac{E^c}{3(1 - 2\nu^c)},\ K^m = \frac{E^m}{3(1 - 2\nu^m)},$$

$$G_z = G^m + \left(G^c - G^m\right)V_c^p\left[1 + \left(1 - V_c^p\right)\left(\frac{G^c - G^m}{G^m + f^m}\right)\right]^{-1},$$

$$G^c = \frac{E^c}{2(1 + \nu^c)},\ G^m = \frac{E^m}{2(1 + \nu^m)},$$

$$f^m = \frac{G^m(9K^m + 8G^m)}{6(K^m + 2G^m)},\ E_z = \frac{9K_zG_z}{3K_z + G_z},$$

$$k_z = k^m + \left(k^c - k^m\right)V_c^p\left[1 + \left(1 - V_c^p\right)\left(\frac{k^c - k^m}{3k^m}\right)\right]^{-1},$$

$$\alpha_z = \alpha^m + \left(\alpha^c - \alpha^m\right)\left(\frac{1}{K_z} - \frac{1}{K^m}\right)\left(\frac{1}{K^c} - \frac{1}{K^m}\right)^{-1}.$$

(iii) **Exponential Function** (Swaminathan and Sangeetha 2017):

$$E_z = E^m e^{\frac{z}{h}\ln\frac{E^c}{E^m}},\ \rho_z = \rho^m e^{\frac{z}{h}\ln\frac{\rho^c}{\rho^m}},\ \alpha_z = \alpha^m e^{\frac{z}{h}\ln\frac{\alpha^c}{\alpha^m}},\ k_z = k^m e^{\frac{z}{h}\ln\frac{k^c}{k^m}}.$$

2.2 *Materials Temperature Dependency*

In many practical applications, particularly, in modern missile technology, space vehicles, aircraft wings, radiant burners, heat exchangers, artillery barrels and micro-electronics, the structural elements have to work under elevated temperature. Such type of heating of structural components leads to the variation in the mechanical properties of the material, i.e. mechanical properties become function of temperature. This necessitates to study the vibration characteristics of beams in the presence of thermal environment. Considering this, the temperature-dependent mechanical properties of the material were developed by Cubberly (1989), Munro (1997), Chan et al. (1991) and Touloukian (1973) and expressed as follows:

$$P^i(T) = P_0^i\left(P_{-1}^i T^{-1} + 1 + P_1^i T^1 + P_2^i T^2 + P_3^i T^3\right),\quad i = m, c, \tag{1.2}$$

where the experimental values of P_j ($j = -1, 0, 1, 2, 3$) refer to any mechanical property of ceramic and metal; T is the temperature within the plate. The experimental values of different ceramics and metals are reported in reference Malekzadeh and Alibeygi Beni (2010).

3 Nanostructures

In the last decade, nanotechnology has become the topic of interest for researchers due to its applications in a variety of complementary areas such as the invention of atomic precise materials; atomic and molecular configurations of materials; the electronic and logic devices with atomic or molecular level materials; modeling and simulation of nanomaterials based on the physics and chemistry. The nanoscale devices and systems are extremely small in size, so exhibit different properties from the conventional ones. Many novel concepts and laws of normal scales are not true for nanoscale structures, and hence modified accordingly and verified by the experiments. Eric Drexler was the first person who introduced the term nanotechnology in 1980s, but it defined by the American Ceramic Society with dimensions of 10^{-9} m. The modern applications of nanostructures such as higher thermal resistance, low weight, mechanical, optical and electrical properties have encouraged the researchers to study their static and dynamic behaviour. As the local continuum theories don't consider the effect of the size scale of the structure, they cannot predict the behaviour of nanostructures. Hence, researchers have proposed various theories such as strain gradient theory, couple stress theory, modified couple stress theory and nonlocal elasticity theory, and acknowledged as nonlocal theories. Out of these, nonlocal elasticity theory (Eringen 2002) is widely used by researchers. According to this theory, the stress tensor at a reference point x in the domain of material is a function of the strains at all points in the neighbourhood of x and can be expressed in the integral form as follows:

$$\sigma_{ij}^{NL} = \int_V \beta(|x' - x|, \tau)\sigma_{ij}^{L}(x')\mathrm{d}V(x'),$$

$$\sigma_{ij}^{L}(x) = C(x):\epsilon_{ij},$$

where σ_{ij}^{L} and σ_{ij}^{NL} are the local and nonlocal stresses, $\beta(|x' - x|, \tau)$ is the nonlocal kernel function, $\tau = e_0c/a$ being the length scale parameter, e_0 is the material constant, and c is an internal characteristic length; $C(x)$ is the elasticity tensor.

An appropriate choice of kernel function (Eringen 1983) leads to the differential form of the nonlocal stress–strain relationship is given by

$$(1 - \mu\nabla^2)\sigma_{ij}^{NL}(x) = C(x):\epsilon_{ij}, \tag{1.3}$$

where the nonlocal parameter $\mu = (e_0c)^2$ controls the size effect for nanostructures; ∇^2 is the Laplacian operator. The nonlocal parameter μ reveals the nonlocal effects, e.g. the effect of internal/cohesive forces on the response of nanostructures, which are prominent for micro/nanostructures (Kröner 1967). By substituting $\mu = 0$, the nonlocal theory reduces to the local theory which is based upon the concept that responses of internal forces have zero range.

In view of the above discussions and classifications, an example is performed to show the formulation of a mathematical model for the vibration analysis of a nonuniform functionally graded nanobeam. The effective temperature-independent and -dependent mechanical properties of the fabricated functionally graded material are calculated by the power law function. The stresses and strains of the beam are obtained in terms of displacement components of the beam using Hooke's law and Euler–Bernoulli's beam theory. Further, the governing equation of the uniform, linear, parabolic, and quadratic nanobeams is derived from Hamilton's principle and Eringen's nonlocal theory. A comparison of results is shown to choose the solution technique, and then the vibration behaviour of nanobeam is investigated for varying values of different parameters for clamped–clamped, clamped–simply supported and simply supported–simply supported boundary conditions.

4 Mathematical Formulation

A functionally graded nanobeam with thickness h and length L is shown in Fig. 2. The upper surface of the beam is fully ceramic, and the lower surface is fully metallic, while there is a mixture of ceramic and metal within the beam. A temperature T is varying in the thickness direction of the beam with T_c and T_m temperature at upper and lower surfaces, respectively. Based on the Euler–Bernoulli beam theory, the displacement components u_x and u_z at the mid-plane ($z = 0$) of the beam can be defined as follows(Bauchau and Craig 2009):

$$u_x(x, z, t) = -z\frac{\partial w}{\partial x}, u_z(x, z, t) = w(x, t), \tag{1.4}$$

According to the above assumptions, the only nonzero strain and stresses of the functionally graded beam are computed by Hooke's law as follows:

$$\epsilon_{xx} = -z\frac{\partial^2 w}{\partial x^2}, \tag{1.5a}$$

$$\sigma_{xx} = E_z(T)\epsilon_{xx}. \tag{1.5b}$$

The strain and kinetic energies of the functionally graded beam are given by

$$U = \frac{1}{2}\int_0^a\int_{-h/2}^{h/2} \sigma_{xx}\epsilon_{xx}\mathrm{d}z\mathrm{d}x, \tag{1.6}$$

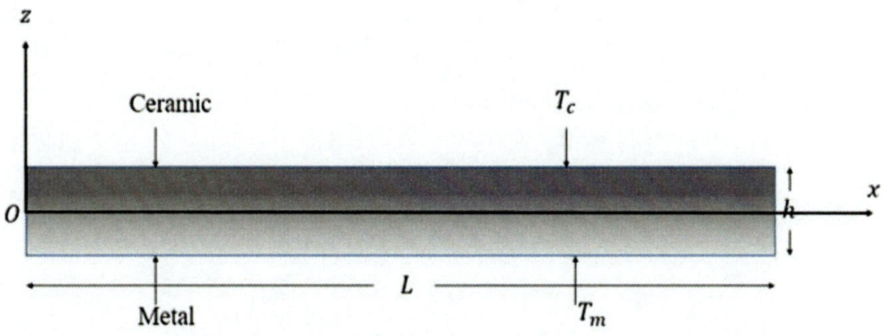

Fig. 2 Geometry of the functionally graded uniform nanobeam subjected to thermal field

$$V = \frac{1}{2}\int_{0}^{L}\int_{-h/2}^{h/2} \rho(z)\left(\frac{\partial u_x}{\partial t}\frac{\partial u_x}{\partial t} + \frac{\partial u_z}{\partial t}\frac{\partial u_z}{\partial t}\right)\mathrm{d}z\mathrm{d}x. \tag{1.7}$$

The work done by the beam due to the thermal stresses is

$$W_T = -\frac{1}{2}\int_{0}^{L}\int_{-h/2}^{h/2} E_z(T)(1+\nu)\alpha_z(T)(T - T_0)\frac{\partial w}{\partial x}\frac{\partial w}{\partial x}\mathrm{d}z\mathrm{d}x. \tag{1.8}$$

To calculate the virtual work done by the beam, taking the variations of energy relations and thermal work

$$\delta U = \int_{0}^{L}\int_{-h/2}^{h/2} \sigma_{xx}\delta\epsilon_{xx}\mathrm{d}z\mathrm{d}x, \tag{1.9}$$

$$\delta V = \int_{0}^{L}\int_{-h/2}^{h/2} \rho_z\left(\frac{\partial u_x}{\partial t}\frac{\partial \delta u_x}{\partial t} + \frac{\partial u_z}{\partial t}\frac{\partial \delta u_z}{\partial t}\right)\mathrm{d}z\mathrm{d}x, \tag{1.10}$$

$$\delta W_T = -\int_{0}^{L}\int_{-h/2}^{h/2} E_z(T)(1+\nu)\alpha_z(T)(T - T_0)\frac{\partial w}{\partial x}\frac{\partial \delta w}{\partial x}\mathrm{d}z\mathrm{d}x. \tag{1.11}$$

Using the strain and stress relations from Eqs. (1.5a, 1.5b) gives

$$\delta U = -\int_{0}^{L} M_{xx}\frac{\partial^2 \delta w}{\partial x^2}\mathrm{d}x, \tag{1.12}$$

$$\delta V = \int_{0}^{L}\left(I_0\frac{\partial w}{\partial t}\frac{\partial \delta w}{\partial t} + I_2\frac{\partial^2 w}{\partial x\partial t}\frac{\partial^2 \delta w}{\partial x\partial t}\right)\mathrm{d}x, \tag{1.13}$$

$$\delta W_T = \int_{0}^{L} N_T\frac{\partial w}{\partial x}\frac{\partial \delta w}{\partial x}\mathrm{d}x, \tag{1.14}$$

where

$$M_{xx} = \int_{-h/2}^{h/2} z\sigma_{xx}\mathrm{d}z,$$

is the normal moment resultant of the beam;

$$[\, I_0 \; I_2 \,] = \int_{-h/2}^{h/2} \rho_z [\, 1 \; z^2 \,] \mathrm{d}z,$$

are the inertia terms of the beam;

$$N_T = -\int_{-h/2}^{h/2} E_z(T)(1+\nu)\alpha_z(T)(T-T_0)\mathrm{d}z,$$

and T_0 is the reference temperature.

The equation of motion has been obtained by Hamilton's principle which can be written as

$$\delta \int_{t_1}^{t_2} L\mathrm{d}t = 0, \tag{1.15}$$

where $L = V - U - W_T$.

Substituting the values for total virtual work L

$$\int_{t_1}^{t_2}\int_0^L \left(I_0 \frac{\partial w}{\partial t}\frac{\partial \delta w}{\partial t} + I_2 \frac{\partial^2 w}{\partial x \partial t}\frac{\partial^2 \delta w}{\partial x \partial t} + M_{xx}\frac{\partial^2 \delta w}{\partial x^2} - N_T \frac{\partial w}{\partial x}\frac{\partial \delta w}{\partial x} \right)\mathrm{d}x\mathrm{d}t = 0. \tag{1.16}$$

Integrating by parts with respect to x and/or t

$$\int_0^L \left(I_0 \frac{\partial w}{\partial t}\delta w + I_2 \frac{\partial^2 w}{\partial x \partial t}\frac{\partial w}{\partial x} - I_2 \frac{\partial^3 w}{\partial x \partial t^2}\frac{\partial \delta w}{\partial x} \right)_{t_1}^{t_2} \mathrm{d}x$$

$$+ \int_{t_1}^{t_2} \left(M_{xx}\frac{\partial \delta w}{\partial x} - \frac{\partial M_{xx}}{\partial x}\delta w - N_T \delta w \right)_0^L \mathrm{d}t$$

$$- \int_{t_1}^{t_2}\int_0^L \left[I_0 \frac{\partial^2 w}{\partial t^2} - I_2 \frac{\partial^4 w}{\partial x^2 \partial t^2} - \frac{\partial^2 M_{xx}}{\partial x^2} - \frac{\partial}{\partial x}\left(N_T \frac{\partial w}{\partial x} \right) \right] \delta w \mathrm{d}x\mathrm{d}t = 0. \tag{1.17}$$

The above integrals will be independently zero, hence, equations can be obtained as

$$\left(I_0\frac{\partial w}{\partial t}\delta w + I_2\frac{\partial^2 w}{\partial x\partial t}\frac{\partial w}{\partial x} - I_2\frac{\partial^3 w}{\partial x\partial t^2}\frac{\partial \delta w}{\partial x}\right)_{t_1}^{t_2} = 0, \tag{1.18a}$$

$$\left(M_{xx}\frac{\partial \delta w}{\partial x} - \frac{\partial M_{xx}}{\partial x}\delta w - N_T\delta w\right)_0^L = 0, \tag{1.18b}$$

$$\frac{\partial^2 M_{xx}}{\partial x^2} + \frac{\partial}{\partial x}\left(N_T\frac{\partial w}{\partial x}\right) - I_0\frac{\partial^2 w}{\partial t^2} + I_2\frac{\partial^4 w}{\partial x^2\partial t^2} = 0. \tag{1.18c}$$

The rotary inertia term with coefficient I_2 is negligible as compared to other inertia terms, hence, neglected (Reddy 2008). However, it can contribute to the higher frequencies. The dynamic part does not contribute to line integrals, so neglected. The temperature is assumed to be constant on the surfaces. Now, Eq. (1.18c) is modified as Euler–Lagrange's equation:

$$\frac{\partial^2 M_{xx}}{\partial x^2} + \frac{\partial}{\partial x}\left(N_T\frac{\partial w}{\partial x}\right) = I_0\frac{\partial^2 w}{\partial t^2}. \tag{1.19}$$

4.1 Boundary Conditions

Primary variables

$$w, \quad \frac{\partial w}{\partial x},$$

Secondary variables

$$M_{xx}, \quad \frac{\partial M_{xx}}{\partial x}.$$

According to the above variables, the boundary conditions of the beam would be

(i) **Clamped (C) edge**:

$$w = \frac{\partial w}{\partial x} = 0,$$

(ii) **Simply supported (S) edge**:

$$w = M_{xx} = 0,$$

(iii) **Free (F) edge**:

$$M_{xx} = \frac{\partial M_{xx}}{\partial x} = 0.$$

4.2 Nonlocal Theory

According to Eringen's nonlocal elasticity theory, the nonlocal moment resultant of the functionally graded nanobeam can be obtained as

$$\left(1 - \mu\nabla^2\right)M_{xx} = -D(T)\frac{\partial^2 w}{\partial x^2}, \tag{1.20}$$

where

$$D = \int_{-h/2}^{h/2} z^2 E_z(T)\mathrm{d}z.$$

The nonlocal governing equation of motion for functionally graded nanobeam under the thermal environment is obtained by solving Eqs. (1.19 and 1.20) and can be written as

$$\begin{aligned}
&\left(D\frac{\partial^4 w}{\partial x^4} + 2\frac{\partial D}{\partial x}\frac{\partial^3 w}{\partial x^3} + \frac{\partial^2 D}{\partial x^2}\frac{\partial^2 w}{\partial x^2} - N_T\frac{\partial^2 w}{\partial x^2} - \frac{\partial N_T}{\partial x}\frac{\partial w}{\partial x}\right)\\
&+\mu\left(N_T\frac{\partial^4 w}{\partial x^4} + 3\frac{\partial N_T}{\partial x}\frac{\partial^3 w}{\partial x^3} + 3\frac{\partial^2 N_T}{\partial x^2}\frac{\partial^2 w}{\partial x^2} + \frac{\partial^3 N_T}{\partial x^3}\frac{\partial w}{\partial x}\right)\\
&= I_0\left(\mu\frac{\partial^4 w}{\partial x^2 \partial t^2} - \frac{\partial^2 w}{\partial t^2}\right).
\end{aligned} \tag{1.21}$$

Assuming that the thickness of the nanobeam is nonuniform and defined as $h = h_0 f(x)$, where

$$f(x) = 1 + \frac{\alpha x}{L} + \frac{\beta x^2}{L}.$$

If $\alpha \neq 0$, $\beta = 0 \Rightarrow$, the thickness of the beam varies linearly; $\alpha = 0$, $\beta \neq 0 \Rightarrow$, the thickness of the beam varies parabolically; $\alpha \neq 0$, $\beta \neq 0 \Rightarrow$, the thickness of the beam varies quadratically. The positive value of these parameters represents that the thickness of the beam increases from length 0 to L, and negative value shows reverse (Fig. 3). Using the harmonic solution along with nondimensional variables $X = x/L$, $H = h_0/L$, $w = LW(X)e^{i\omega t}$, $\gamma = \sqrt{\mu}/L$, and $\Omega^2 = \omega^2 L^4 \rho_0^c H/D^*$, Eq. (1.21) is transformed to the nondimensional ordinary differential equation as

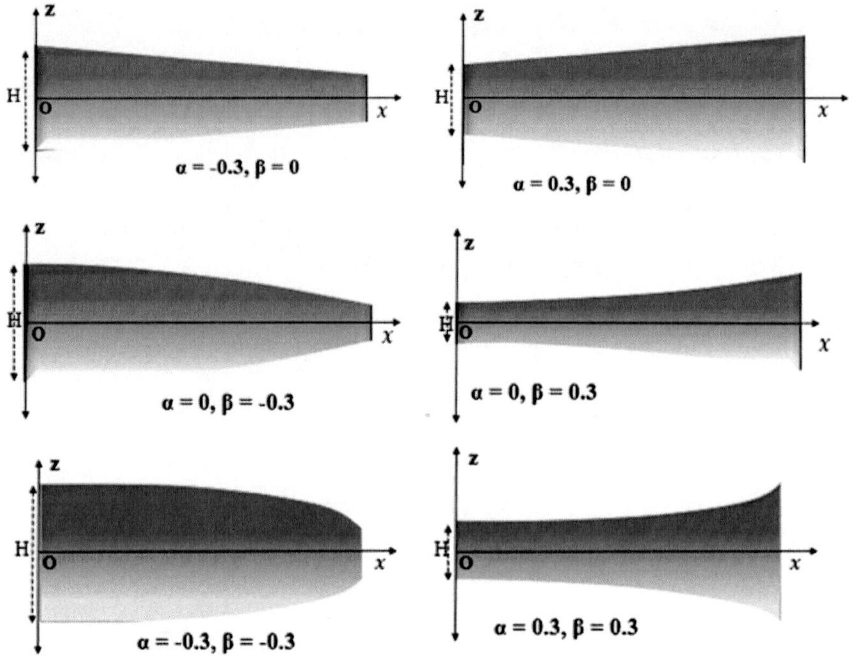

Fig. 3 Representation of linear, parabolic and quadratic thickness variations

$$\left\{B\frac{\mathrm{d}^4W}{\mathrm{d}X^4}+2\frac{\mathrm{d}B}{\mathrm{d}X}\frac{\mathrm{d}^3W}{\mathrm{d}X^3}-\overline{N}_T\frac{\mathrm{d}^2W}{\mathrm{d}X^2}+\frac{\mathrm{d}^2B}{\mathrm{d}X^2}\frac{\mathrm{d}^2W}{\mathrm{d}X^2}-\frac{\mathrm{d}\overline{N}_T}{\mathrm{d}X}\frac{\mathrm{d}W}{\mathrm{d}X}\right\}$$
$$+\gamma^2\left\{\overline{N}_T\frac{\mathrm{d}^4W}{\mathrm{d}X^4}+3\frac{\mathrm{d}\overline{N}_T}{\mathrm{d}X}\frac{\mathrm{d}^3W}{\mathrm{d}X^3}+3\frac{\mathrm{d}^2\overline{N}_T}{\mathrm{d}X^2}\frac{\mathrm{d}^2W}{\mathrm{d}X^2}+\frac{\mathrm{d}^3\overline{N}_T}{\mathrm{d}X^3}\frac{\mathrm{d}W}{\mathrm{d}X}\right\}$$
$$=\Omega^2\overline{I}_0\left(W-\gamma^2\frac{\mathrm{d}^2W}{\mathrm{d}X^2}\right), \tag{1.22}$$

$$\left[\left(B+\gamma^2\overline{N}_T\right)\frac{\mathrm{d}^4W}{\mathrm{d}X^4}+\left(2\frac{\mathrm{d}B}{\mathrm{d}X}+3\gamma^2\frac{\mathrm{d}\overline{N}_T}{\mathrm{d}X}\right)\frac{\mathrm{d}^3W}{\mathrm{d}X^3}\right.$$
$$\left.+\left(\frac{\mathrm{d}^2B}{\mathrm{d}X^2}-\overline{N}_T+3\gamma^2\frac{\mathrm{d}^2\overline{N}_T}{\mathrm{d}X^2}\right)\frac{\mathrm{d}^2W}{\mathrm{d}X^2}-\left(\frac{\mathrm{d}\overline{N}_T}{\mathrm{d}X}-\gamma^2\frac{\mathrm{d}^3\overline{N}_T}{\mathrm{d}X^3}\right)\frac{\mathrm{d}W}{\mathrm{d}X}\right]$$
$$=\Omega^2\overline{I}_0\left(W-\gamma^2\frac{\mathrm{d}^2W}{\mathrm{d}X^2}\right), \tag{1.23}$$

where

$$D=H^3BD^*,\quad I_0=H\overline{I}_0\rho_0^c,\quad N_T=H\overline{N}_TD^*,\quad D^*=\frac{E_0^cH^3}{12}.$$

5 Temperature Profile

The distribution of the temperature within the beam is based on the various models proposed by the researchers such as constant, linear, parabolic, nonlinear and sinusoidal. The direction of the variation of the temperature is based upon the non-homogeneity of the beam material either in thickness and/or in-plane (Saini and Lal 2020). Since most of the material structures are nonhomogeneous in thickness direction rather than in-plane, the corresponding temperature also varies in the thickness direction. Accordingly, the models are as follows.

5.1 *Constant and Linear Variation*

If T_c and T_m are the temperatures at the upper and lower surfaces, respectively, then the expression for the linear temperature variation is given by

$$T(z) = T_m + (T_c - T_m)V_c$$

However, for uniform variation, $T_c = T_m$.

5.2 *Nonlinear Variation*

Generally, nonlinear temperature variation is obtained by solving a three-dimensional heat conduction equation (Swaminathan and Sangeetha 2017). Due to the non-homogeneity in onedirection, most of the studies are focused on one-dimensional temperature variation. Consequently, the one-dimensional equation in the absence of heat flux is

$$\frac{\mathrm{d}}{\mathrm{d}z}\left(k_z \frac{\mathrm{d}T}{\mathrm{d}z}\right) = 0. \tag{1.24}$$

By solving, the temperature can be obtained as (Lal and Saini 2020)

$$T(z) = T_m + \frac{\Delta T}{C^*}\sum_{i=0}^{14}(-1)^i \frac{k_{cm}^i}{(ip+1)}\left(\frac{z}{h} + \frac{1}{2}\right)^{ip+1},$$

where

$$k_{cm} = \frac{(k_c - k_m)}{k_m},\ \Delta T = T_c - T_m,\ C^* = \sum_{i=0}^{14}(-1)^i \frac{k_{cm}^i}{(ip+1)}.$$

6 Methods

Equation (1.23) is a fourth-order ordinary differential equation with variable coefficients of the derivatives. Hence, their analytical solution is not possible except for specific values of various parameters. The approximate solutions are obtained for Eq. (1.23) along with four boundary conditions at the edges.

6.1 *Generalized Differential Quadrature Method*

According to the generalized differential quadrature (GDQ) (Shu 2000) method, the domain [0, 1] of the plate is discretized into x_1, x_2, ..., x_n points. The mth order derivative of a function $W(X)$ at point X_i is approximated by

$$\left\{\frac{\mathrm{d}^m W(X)}{\mathrm{d}X^m}\right\}_{X=X_i} = \sum_{j=1}^{n} C_{ij}^{(m)} W_j, \quad i = 1, 2, \ldots, n, \tag{1.25}$$

where $C_{ij}^{(m)}$ are the weighting coefficients which can be obtained as

$$C_{ij}^{(1)} = \begin{cases} \frac{M^{(1)}(X_i)}{M^{(1)}(X_j)(X_i - X_j)}, & i \neq j \\ -\sum\limits_{j=1 i\neq j}^{n} C_{ij}^{(1)}, & i = j, \end{cases} \quad \text{where } i, j = 1, 2, \ldots, n \text{ and } M^{(1)}(X_i) = \prod_{k=1, k\neq i}^{n} (X_i - X_k),$$

$$C_{ij}^{(m)} = \begin{cases} m\left[C_{ij}^{(1)} C_{ij}^{(m-1)} - \frac{C_{ij}^{(m-1)}}{(X_i - X_j)}\right], & i \neq j \\ -\sum\limits_{j=1, i\neq j}^{n} C_{ij}^{(m)}, & i = j \end{cases} \quad \text{where, } m \geq 2, \quad i, j = 1, 2, \ldots, n.$$

6.2 *Differential Quadrature Element Methods*

According to the differential quadrature element methods (DQEM) (Wang 2015), each endpoint of the beam has two degrees of freedom which is equal to the number of boundary conditions. So, extending the GDQ method by introducing $W_1^{(1)} = \left(\frac{\mathrm{d}W}{\mathrm{d}X}\right)_{X=X_1}$ and $W_n^{(1)} = \left(\frac{\mathrm{d}W}{\mathrm{d}X}\right)_{X=X_n}$ corresponding to the second degree of freedom at nodes $X = X_1$ and $X = X_n$, respectively, which leads to $n + 2$ unknowns during the implementation of DQEM as compared to n unknowns in the GDQ method. Now Eq. (1.25) reduces to

$$\left\{\frac{\mathrm{d}^m W(X)}{\mathrm{d}X^m}\right\}_{X=X_i} = \sum_{j=1}^{n} h_{j0}^{(m)}(X_i)W_j + h_{11}^{(m)}(X_i)W_1^{(1)} + h_{n1}^{(m)}(X_i)W_n^{(1)}$$

$$= \sum_{j=1}^{n+2} E_{j0}^{(m)}(X_i)W_j, \quad (i = 1, 2, \ldots, n). \tag{1.26}$$

Further, two types of DQEM are available in the literature based on the choice of interpolating functions h_{ij} as follows:

New Version of Differential Quadrature Method (NDQ)

In this approach, Lagrange's interpolation functions are used as the basis polynomials. The incorporation of two new variables leads to the modification in weighting coefficients, i.e. $E_{ij}^{(m)}$'s which are as follows:

$$E_{ij}^{(1)} = \begin{cases} C_{ij}^{(1)} & i, j = 1, 2, \ldots, n \\ 0 & i = 1, 2, \ldots, n, \quad j = n+1, n+2, \end{cases}$$

$$E_{ij}^{(2)} = \begin{cases} C_{ij}^{(2)} & i = 2, 3, \ldots, n-1, j = 1, 2, \ldots, n, \\ 0 & i = 1, 2, \ldots, n, j = n+1, n+2, \end{cases}$$

$$E_{ij}^{(2)} = \sum_{k=2}^{n-1} C_{ik}^{(1)} C_{kj}^{(1)}, E_{i(n+1)}^{(2)} = C_{i1}^{(1)}, E_{i(n+2)}^{(2)} = C_{in}^{(1)}, \quad i = 1, n,$$

$$E_{ij}^{(3)} = \sum_{k=1}^{n} C_{ik}^{(1)} E_{kj}^{(2)}, \quad i = 1, 2, 3, \ldots n, \quad j = 1, 2, \ldots n+2,$$

$$E_{ij}^{(4)} = \sum_{k=1}^{n} C_{ik}^{(2)} E_{kj}^{(2)}, \quad i = 1, 2, 3, \ldots n, \quad j = 1, 2, \ldots n+2.$$

Generalized Differential Quadrature Rule

The Hermite interpolation functions are used in this approach which are as follows:

$$h_{jk}(X) = \left(a_{jk}X^2 + b_{jk}X + c_{jk}\right)l_j(X), \quad j = 1, 2, \ldots, n, \quad k = 0, 1$$

where $l_j(X)$ are the Lagrange's interpolation functions, defined as

$$l_j(X) = \prod_{k=1,k\neq j}^{n} \frac{(X - X_k)}{(X_j - X_k)},$$

$$a_{jk} = \begin{cases} \frac{1}{X_j^2 - X_j(X_1+X_n)+X_1X_n}, & j = 2, 3, \ldots, n-1, \ k = 0, \\ \frac{-1}{(X_j - X_{n+1-j})^2} - \frac{l_j^1(X_j)}{(X_j - X_{n+1-j})}, & j = 1, n, \quad k = 0, \\ \frac{1}{X_j - X_{n+1-j}}, & j = 1, n, \quad k = 1, \end{cases}$$

$$b_{jk} = \begin{cases} \frac{-(X_1+X_n)}{X_j^2 - X_j(X_1+X_n)+X_1X_n}, & j = 2, 3, \ldots, n-1, k = 0, \\ \frac{1}{X_j - X_{n+1-j}} - a_{jk}\left(X_j + X_{n+1-j}\right), & j = 1, n, k = 0, \\ -\frac{X_j + X_{n+1-j}}{X_j - X_{n+1-j}}, & j = 1, n, \quad k = 1, \end{cases}$$

$$c_{jk} = \begin{cases} \frac{X_1X_n}{X_j^2 - X_j(X_1+X_n)+X_1X_n}, & j = 2, 3, \ldots, n-1, k = 0, \\ 1 - a_{jk}X_j^2 - b_{jk}X_j, & j = 1, n, k = 0, \\ \frac{X_jX_{n+1-j}}{X_j - X_{n+1-j}}, & j = 1, n, k = 1. \end{cases}$$

6.3 Grid Points Distribution

The convergence and accuracy of the method also depend on the choice of grid points distribution. Consequently, Chebyshev Gauss–Lobatto grid distribution (Wang 2015) is considered for the analysis:

$$X_i = \frac{1}{2}\left[1 - \cos\left(\frac{i-1}{n-1}\pi\right)\right], \quad i = 1, 2, \ldots, n. \tag{1.27}$$

6.4 Solution Procedure

Equation (1.23) is discretized by approximating the derivatives of nondimensional transverse displacement W using the above-mentioned methods at any grid point x_i, $i = 1, 2, \ldots, n$ gives simultaneous equations in terms of unknowns W_is. The satisfaction of the resultant equation at grid points all grid points provides a set of linear equations. These equations together with corresponding boundary conditions provide an eigenvalue problem which is solved to compute the non-dimensional frequencies of the nanobeam using MATLAB.

7 Results and Discussion

Thermoelastic vibrations of functionally graded nonuniform nanobeams with CC, CS, SS and CF supports are investigated to discuss the effect of size dependency, volume fraction index, taper parameters and temperature difference. The mechanical properties are listed in Table 1 for ceramic and metal constituents as Si_3N_4 and SuS304, respectively. The values of various parameters are taken as follows: nonlocal parameter $\gamma = 0.0, 0.1, 0.2, 0.3$, temperature difference $\Delta T = 0(20)200$ K, volume fraction index $p = 0, 0.5, 2$, and taper parameters $\alpha, \beta = -0.3, 0, 0.3$.

7.1 Convergence and Comparison

To choose a suitable method from the above methods, a comparison test is made in view of the required grid points for the accuracy of four decimal places in the frequency parameter of nonuniform nanobeams with different boundary conditions. In view of this, the results are reported in Table 2 considering parameter $p = 2$, $\Delta T = 200$ K, $\gamma = 0.2$. It can be seen that the NDQ method needs less number of grid points as compared to other methods, so this method has a faster rate of convergence. Accordingly, 22 grid points are fixed for the desired accuracy. Further, the comparison of results with those obtained by other methods is presented in Table 3 which justifies that the present results are trustworthy.

Table 1 Mechanical properties for ceramic and metal

TD	E (Pa)		ρ (kg/m^3)		α (1/K)		k (W/mK)	
	SuS304	Si_3N_4	SuS304	Si_3N_4	SuS304	Si_3N_4	SuS304	Si_3N_4
P_{-1}	0	0	0	0	0	0	0	0
P_0	201.04×10^9	348.43×10^9	8166	2370	12.33×10^{-6}	5.87×10^{-6}	12.04	9.19
P_1	3.079×10^{-4}	-3.07×10^{-4}	0	0	8.086×10^{-4}	9.095×10^{-4}	0	0
P_2	-6.534×10^{-7}	2.16×10^{-7}	0	0	0	0	0	0
P_3	0	-8.946×10^{-11}	0	0	0	0	0	0
TI	Aluminium	Al_2O_3	Aluminium	Al_2O_3	Aluminium	Al_2O_3	Aluminium	Al_2O_3
	70×10^9	380×10^9	2707	3800	23×10^{-6}	7.4×10^{-6}	204	10.4

Table 2 Convergence of frequency parameter $p = 2$, $\Delta T = 200$ K, $\gamma = 0.2$

			GDQ method			NDQ method			GDQ rule		
			I	II	III	I	II	III	I	II	III
CC	Uniform	(0, 0)	11	15	17	12	14	17	40	48	55
	Linear	(−0.3, 0)	11	14	17	11	16	17	42	49	58
		(0.3, 0)	11	16	17	10	14	16	41	47	57
	Parabolic	(0, −0.3)	11	12	15	10	12	15	43	50	58
		(0, 0.3)	12	15	19	12	13	17	42	48	59
	Quadratic	(−0.3, −0.3)	15	18	19	15	18	19	44	51	60
		(0.3, 0.3)	13	16	19	13	16	17	47	53	63
CS	Uniform	(0, 0)	11	14	19	10	13	16	40	48	55
	Linear	(−0.3, 0)	11	14	17	11	12	15	42	49	58
		(0.3, 0)	12	15	18	10	13	15	41	47	57
	Parabolic	(0, −0.3)	12	13	15	10	11	13	43	50	58
		(0, 0.3)	14	17	19	11	14	17	42	48	59
	Quadratic	(−0.3, −0.3)	20	20	21	15	15	18	44	51	60
		(0.3, 0.3)	13	16	20	11	13	16	47	53	63
SS	Uniform	(0, 0)	11	14	17	8	10	13	41	49	56
	Linear	(−0.3, 0)	11	14	17	8	10	13	42	48	57
		(0.3, 0)	11	14	17	8	10	12	40	46	55
	Parabolic	(0, −0.3)	13	14	14	9	9	11	41	51	56
		(0, 0.3)	13	16	17	8	11	12	43	49	57
	Quadratic	(−0.3, −0.3)	21	20	21	15	14	14	45	50	59
		(0.3, 0.3)	13	17	18	9	10	13	46	51	61
CF	Uniform	(0, 0)	10	15	18	8	11	13	46	56	68
	Linear	(−0.3, 0)	12	13	16	9	11	14	48	59	71
		(0.3, 0)	11	14	18	10	11	14	47	58	69
	Parabolic	(0, −0.3)	15	15	16	10	11	13	49	61	73
		(0, 0.3)	13	16	20	10	13	15	48	59	71
	Quadratic	(−0.3, −0.3)	22	23	25	18	18	18	51	63	74
		(0.3, 0.3)	11	16	23	9	11	16	52	64	73

Table 3 Comparison of frequency parameter for isotropic simply supported nanobeam

$\gamma^2 \rightarrow$	0	0.01	0.02	0.03	0.04
Present	9.8696	9.4159	9.0195	8.6693	8.3569
Reddy (2007)	9.8696	9.4159	9.0195	8.6693	8.3569
Aydogdu (2009)	9.8696	9.4124	9.0133	8.6611	8.3462
Eltaher et al. (2012)	9.8696	9.4159	9.0195	8.6693	8.3569

7.2 Parametric Discussion

In Tables 4, 5 and 6, the values of frequency parameter of nonuniform nanobeam are reported for the first three modes of vibration considering temperature-independent (TI) and temperature-dependent (TD) mechanical properties. The choice of the number of modes is based on the fact that initial modes have most of the energies of the structure. It can be observed that the value of Ω increases with the increase in the number of modes of the beam, keeping other parameters fixed for all combinations of boundary conditions. Also, the values of frequency parameter for nonuniform nanobeams of TI materials are found to be higher as compared to those for the TD materials. Further, due to the edge fixity of the nanobeams, the values of Ω are found in the order of $\Omega_{CC} > \Omega_{CS} > \Omega_{SS} > \Omega_{CF}$.

Table 4 represents the values of fundamental frequency Ω of functionally graded nonuniform nanobeams with $\gamma = 0, 0.2,\ p = 0, 2$, and $\Delta T = 0, 200$ K for CC, CS, SS and CF supports at the edges of the beam. The values of frequency parameter are found to decrease with the large values of volume fraction index and temperature difference. The increase in the fraction index leads to a higher volume of metallic constituent, and hence, frequency parameter decreases due to lower stiffness. Further, the temperature has the softening effect, and accordingly, frequency decreases. However, for the nonlocal parameter, the behaviour of frequency is based on the edge support, i.e. it decreases for CC, CS and SS boundary conditions but increases for CF boundary conditions. Various researchers observed and reported such paradoxical behaviour of nanobeams for CF boundary conditions. In view of different thickness variations considered in the analysis, the order of the values of frequency parameter for the first mode is reported in Table 7.

In Tables 5 and 6, the values of frequency parameter are reported for modes second and third, respectively. Similar to the first mode, the value of frequency parameter decreases with higher values of volume fraction index and temperature difference for the same set of the values of other parameters. Also, it decreases with the increasing values of nonlocal parameter irrespective of the boundary conditions. The changes in the values of Ω are also observed in view of the reported values of parameters, and it is observed that the volume fraction index affects most while the temperature difference has the lowest impact on the frequency. This effect further increases when the same nonuniform beam is fabricated with TI materials. Further, the values of frequency parameter in the second and third modes for nonuniform nanobeam are found to be in the order of $\Omega_{\text{Uniform}} > \Omega_{\text{Parabolic}} > \Omega_{\text{Linear}} > \Omega_{\text{Quadratic}}$ when the thickness of the beam is decreasing from left to right with CC, CS and SS supports, while it is reverse for the increment in the thickness from left to right.

As the nonuniform functionally graded nanobeam with CF edges shows paradoxical behaviour in the fundamental mode, the corresponding graphs of frequency parameter Ω are plotted in Figs. 4 and 5 for varying values of temperature difference ΔT taking $p = 0.5, 2$, $\gamma = 0.1$ along with TD and TI material properties. Figure 4 illustrates the graphs for the nanobeam considering uniform, linear, parabolic and quadratic thickness variations such that the thickness decreases from left to right. It

Table 4 Numerical values of frequency parameter of the first mode of nonuniform nanobeam

Thickness variation →				Uniform		Linear				Parabolic				Quadratic			
	(α, β) →			(0, 0)		(−0.3, 0)		(0.3, 0)		(0, −0.3)		(0, 0.3)		(−0.3, −0.3)		(0.3, 0.3)	
	ΔT↓	p↓	γ↓	TD	TI	TD	TI	TD	TI	TD	TI	TD	TI	TD	TI	TD	TI
CC	0	0	0	21.5451	22.3733	19.7861	20.5466	23.0555	23.9417	19.5757	20.3281	23.2757	24.1704	17.3556	18.0227	24.6065	25.5524
			0.2	17.6124	18.2894	16.2064	16.8294	18.8673	19.5925	16.0458	16.6626	19.0396	19.7714	14.3718	14.9242	20.1788	20.9544
		2	0	11.7763	17.7814	10.8148	16.3297	12.6019	19.0280	10.6998	16.1560	12.7222	19.2097	9.4863	14.3237	13.4496	20.3081
			0.2	9.6267	14.5357	8.8583	13.3754	10.3126	15.5714	8.7704	13.2428	10.4068	15.7136	7.8555	11.8612	11.0295	16.6538
	200	0	0	21.0422	22.0534	19.1977	20.1327	22.6035	23.6813	19.0333	19.9564	22.7979	23.8871	16.6790	17.5073	24.1681	25.3162
			0.2	16.7403	17.5903	15.1061	15.9042	18.1189	19.0199	15.0409	15.8265	18.2390	19.1506	12.8697	13.6106	19.4685	20.4277
		2	0	11.4057	17.3779	10.3716	15.8064	12.2752	18.6998	10.2937	15.6864	12.3750	18.8526	8.9662	13.6701	13.1366	20.0107
			0.2	8.9472	13.6474	7.9856	12.1917	9.7371	14.8457	7.9769	12.1751	9.7888	14.9261	6.6178	10.1283	10.4874	15.9867
CS	0	0	0	14.8475	15.4182	14.4553	15.0109	15.1694	15.7524	14.5247	15.0830	15.0322	15.6100	13.9013	14.4356	15.2975	15.8855
			0.2	12.2740	12.7458	11.9553	12.4148	12.5676	13.0507	12.0351	12.4977	12.4415	12.9197	11.6304	12.0774	12.7041	13.1924
		2	0	8.1155	12.2538	7.9011	11.9301	8.2914	12.5194	7.9390	11.9874	8.2164	12.4063	7.5983	11.4729	8.3614	12.6252
			0.2	6.7088	10.1299	6.5346	9.8668	6.8693	10.3722	6.5782	9.9327	6.8004	10.2681	6.3570	9.5987	6.9439	10.4849
	200	0	0	14.2722	14.9806	13.7505	14.4477	14.6763	15.3953	13.8903	14.5870	14.5012	15.2157	13.0835	13.7616	14.8336	15.5569
			0.2	11.4708	12.0735	10.9047	11.5068	11.9012	12.5101	11.0762	11.6768	11.7304	12.3359	10.1442	10.7615	12.0931	12.7054
		2	0	7.6736	11.6992	7.3518	11.2136	7.9175	12.0677	7.4479	11.3576	7.8112	11.9072	6.9536	10.6114	8.0115	12.2100
			0.2	6.0736	9.2714	5.6899	8.6963	6.3481	9.6846	5.8111	8.8775	6.2420	9.5245	5.1124	7.8388	6.4686	9.8661
SS	0	0	0	9.5043	9.8696	8.6784	9.0120	10.139	10.5287	9.0067	9.3529	9.8411	10.2193	7.8568	8.1587	10.3752	10.7740
			0.2	8.0476	8.3569	7.3604	7.6434	8.5927	8.9230	7.6389	7.9325	8.3412	8.6619	6.7168	6.9750	8.8118	9.1505
		2	0	5.1949	7.8440	4.7435	7.1624	5.5418	8.3678	4.9230	7.4333	5.3790	8.1219	4.2944	6.4843	5.6710	8.5628

(continued)

Table 4 (continued)

Thickness variation →				Uniform		Linear				Parabolic				Quadratic			
		(α, β) →		(0, 0)		(−0.3, 0)		(0.3, 0)		(0, −0.3)		(0, 0.3)		(−0.3, −0.3)		(0.3, 0.3)	
	ΔT↓	p↓	γ↓	TD	TI	TD	TI	TD	TI	TD	TI	TD	TI	TD	TI	TD	TI
			0.2	4.3987	6.6418	4.0231	6.0747	4.6967	7.0917	4.1753	6.3044	4.5592	6.8841	3.6713	5.5435	4.8164	7.2724
	200	0	0	8.8035	9.2745	7.7920	8.2359	9.5539	10.0475	8.2346	8.6869	9.1938	9.6772	6.8228	7.2384	9.8244	10.3271
			0.2	7.2338	7.6449	6.3039	6.6987	7.9181	8.3461	6.6894	7.0902	7.6118	8.0315	5.2919	5.6860	8.1889	8.6241
		2	0	4.6375	7.0822	4.0268	6.1595	5.0822	7.7550	4.3042	6.5775	4.8671	7.4297	3.4460	5.2817	5.2402	7.9943
			0.2	3.7411	5.7221	3.1510	4.8342	4.1583	6.3520	3.3987	5.2065	3.9740	6.0734	2.4383	3.7719	4.3218	6.5990
CF	0	0	0	3.3863	3.5155	4.0543	4.2108	2.9535	3.0666	4.1111	4.2700	2.3878	2.9982	5.2620	5.4637	2.5900	2.6807
			0.2	3.4473	3.5789	4.1710	4.3319	2.9906	3.1050	4.2243	4.3876	2.9243	3.0360	5.5272	5.7391	2.6141	2.7057
		2	0	1.8505	2.7930	2.2161	3.3449	1.6140	2.4357	2.2469	3.3917	1.5782	2.3804	2.8761	4.3428	1.4160	2.1343
			0.2	1.8839	2.8434	2.2798	3.4412	1.6342	2.4662	2.3088	3.4852	1.5981	2.4104	3.0211	4.5617	1.4292	2.1542
	200	0	0	3.5128	3.6608	4.1050	4.2875	3.1092	3.2331	4.0828	4.2720	3.1010	3.2213	4.9838	5.2364	2.8037	2.9065
			0.2	3.5011	3.6531	4.1098	4.3018	3.0932	3.2222	4.0970	4.2929	3.0785	3.2030	5.0795	5.3516	2.7823	2.8821
		2	0	1.9606	2.9809	2.2682	3.4495	1.7457	2.6512	2.2352	3.4028	1.7537	2.6629	2.6672	4.0673	1.5904	2.4051
			0.2	1.9376	2.9476	2.2462	3.4187	1.7248	2.6235	2.2198	3.3817	1.7297	2.6273	2.6843	4.0970	1.5691	2.3751

Table 5 Numerical values of frequency parameter of the second mode of nonuniform nanobeam

Thickness variation →				Uniform		Linear				Parabolic				Quadratic			
		(α, β) →		(0, 0)		(−0.3, 0)		(0.3, 0)		(0, −0.3)		(0, 0.3)		(−0.3, −0.3)		(0.3, 0.3)	
	$\Delta T\downarrow$	$p\downarrow$	$\gamma\downarrow$	TD	TI	TD	TI	TD	TI	TD	TI	TD	TI	TD	TI	TD	TI
CC	0	0	0	59.3900	61.6728	54.5586	56.6557	63.5645	66.0078	55.0579	57.1742	63.1348	65.5616	49.1399	51.0287	66.8918	69.4630
			0.2	35.0757	36.4239	32.2139	33.4521	37.5358	38.9786	32.5777	33.8299	37.2310	38.6621	29.0855	30.2034	39.4349	40.9507
		2	0	32.4619	49.0153	29.8211	45.0278	34.7436	52.4605	30.0940	45.4399	34.5087	52.1059	26.8593	40.5557	36.5623	55.2066
			0.2	19.1720	28.9484	17.6077	26.5865	20.5166	30.9787	17.8066	26.8868	20.3500	30.7272	15.8978	24.0046	21.5547	32.5461
	200	0	0	58.4761	61.2402	53.5465	56.0942	62.7016	65.6544	54.0953	56.6634	62.2466	65.1809	47.9981	50.3042	66.0386	69.1433
			0.2	33.6286	35.3064	30.4061	31.9723	36.2807	38.0613	30.9091	32.4846	35.9011	37.6716	26.5897	28.0574	38.2430	40.1075
		2	0	31.8231	48.4708	29.0947	44.3205	34.1530	52.0160	29.4085	44.7967	33.8970	51.6270	26.0183	39.6416	35.9851	54.8045
			0.2	18.0566	27.5319	16.1882	24.7001	19.5626	29.8183	16.5028	25.1743	19.3344	29.4733	13.8591	21.1857	20.6554	31.4797
CS	0	0	0	48.1154	49.9649	45.0321	46.7631	50.8530	52.8077	45.8305	47.5921	50.1571	52.0850	42.2567	43.8810	52.6999	54.7256
			0.2	29.2140	30.3370	27.2126	28.2585	30.9882	32.1793	27.7158	28.7811	30.5671	31.7421	25.4149	26.3918	32.2132	33.4514
		2	0	26.2993	39.7102	24.6140	37.1655	27.7957	41.9696	25.0504	37.8244	27.4153	41.3952	23.0970	34.8750	28.8051	43.4938
			0.2	15.9680	24.1107	14.8741	22.4588	16.9378	25.5749	15.1491	22.8742	16.7076	25.2274	13.8915	20.9752	17.6074	26.5859
	200	0	0	47.2199	49.4668	43.9902	46.1033	50.0388	52.4071	44.8467	46.9934	49.3159	51.6536	41.0154	43.0115	51.9163	54.3677
			0.2	27.9184	29.3205	25.5520	26.8821	29.8832	31.3567	26.1725	27.5194	29.4026	30.8599	22.9977	24.2947	31.1808	32.7067
		2	0	25.6563	39.0829	23.8471	36.3333	27.2231	41.4653	24.3319	37.0696	26.8197	40.8522	22.1627	33.7757	28.2604	43.0435
			0.2	14.9650	22.8211	13.5643	20.7016	16.0941	24.5337	13.9376	21.2658	15.8144	24.1099	11.9035	18.2076	16.8248	25.6437
SS	0	0	0	38.0171	39.4784	35.0108	36.3566	40.7441	42.3103	36.0748	37.4614	39.7684	41.2970	32.6933	33.9499	42.3556	43.9837
			0.2	23.6724	24.5823	21.8144	22.6529	25.3791	26.3546	22.4852	23.3495	24.7735	25.7257	20.4841	21.2714	26.4026	27.4175
		2	0	20.7797	31.3760	19.1365	28.8948	22.2703	33.6266	19.7180	29.7729	21.7369	32.8213	17.8698	26.9821	23.1511	34.9566

(continued)

Table 5 (continued)

Thickness variation →			Uniform		Linear				Parabolic				Quadratic			
	(α, β) →		(0, 0)		(−0.3, 0)		(0.3, 0)		(0, −0.3)		(0, 0.3)		(−0.3, −0.3)		(0.3, 0.3)	
	$\Delta T\downarrow$	$p\downarrow$	$\gamma\downarrow$	TD	TI	TD	TI	TD	TI	TD	TI	TD	TI	TD	TI	TD	TI
			0.2	12.9390	19.5371	11.9235	18.0037	13.8719	20.9456	12.2902	18.5573	13.5409	20.4458	11.1963	16.9057	14.4314	21.7904
	200	0	0	37.1118	38.8969	33.9462	35.6017	39.9305	41.8360	35.0795	36.7803	38.9206	40.7830	31.4333	32.9928	41.5755	43.5530
			0.2	22.4944	23.6372	20.3213	21.3962	24.3662	25.5786	21.0915	22.1917	23.7106	24.8978	18.3420	19.3990	25.4539	26.7098
		2	0	20.1114	30.6426	18.3333	27.9409	21.6810	33.0290	18.9733	28.9128	21.1183	32.1735	16.9023	25.7690	22.5917	34.4142
			0.2	12.0210	18.3361	10.7389	16.3956	13.0927	19.9621	11.1900	17.0787	12.7195	19.3956	9.4264	14.4277	13.7063	20.8940
CF	0	0	0	21.2189	22.0344	21.2079	22.0232	21.3526	22.1733	21.8891	22.7307	20.8413	21.6423	22.5158	23.3810	21.0612	21.8698
			0.2	16.9255	17.5759	16.8865	17.5357	17.0818	17.7383	17.4718	18.1436	16.6555	17.2956	18.0418	18.7351	16.8927	17.5411
		2	0	11.5980	17.5120	11.5920	17.5029	11.6710	17.6223	11.9643	18.0651	11.3916	17.2001	12.3069	18.5825	11.5118	17.3818
			0.2	9.2512	13.9685	9.2299	13.9364	9.3367	14.0976	9.5499	14.4194	9.1036	13.7455	9.8614	14.8901	9.2334	13.9415
	200	0	0	20.6839	21.6818	20.5828	21.5849	20.8752	21.8759	21.2953	22.3271	20.3400	21.3188	21.7348	22.8043	20.6069	21.5930
			0.2	16.0103	16.8311	15.6899	16.5248	16.3187	17.1387	16.3611	17.2172	15.8522	16.6550	16.3481	17.2651	16.1973	17.0041
		2	0	11.2002	17.0663	11.1203	16.9473	11.3206	17.2472	11.5198	17.5545	11.0205	16.7914	11.7107	17.8509	11.1794	17.0308
			0.2	8.5335	13.0194	8.2763	12.6377	8.7450	13.3363	8.6705	13.2345	8.4778	12.9307	8.4823	12.9710	8.6966	13.2593

Table 6 Numerical values of frequency parameter of the third mode of nonuniform nanobeam

Thickness variation→			Uniform		Linear				Parabolic				Quadratic			
	(α, β)→		(0, 0)		(−0.3, 0)		(0.3, 0)		(0, −0.3)		(0, 0.3)		(−0.3, −0.3)		(0.3, 0.3)	
	ΔT↓	p↓	γ ↓	TD	TI	TD	TI	TD	TI	TD	TI	TD	TI	TD	TI	TD	TI
CC	0	0	0	116.4281	120.9034	106.9714	111.0831	124.6211	129.4113	108.884	113.0693	122.9287	127.6538	97.4618	101.208	130.365	135.376
			0.2	52.5058	54.5240	48.2240	50.0776	56.1897	58.3495	49.1883	51.0790	55.3400	57.4671	43.9999	45.6912	58.6701	60.9252
		2	0	63.6382	96.0895	58.4693	88.2847	68.1164	102.8512	59.5147	89.8632	67.1914	101.4545	53.2715	80.4364	71.2560	107.5917
			0.2	28.6990	43.3336	26.3586	39.7998	30.7126	46.3740	26.8857	40.5957	30.2482	45.6727	24.0498	36.3137	32.0684	48.4211
	200	0	0	115.0308	120.4303	105.498	110.4681	123.2508	129.0241	107.4538	112.5094	121.5402	127.237	95.8642	100.4049	128.9881	135.0248
			0.2	50.5353	53.0369	45.7698	48.1011	54.4678	57.1255	46.9112	49.2774	53.5251	56.1486	40.5412	42.7439	57.0288	59.7960
		2	0	62.7044	95.4947	57.4583	87.5111	67.2179	102.3646	58.5417	89.1591	66.2754	100.9305	52.1440	79.4253	70.3617	107.1504
			0.2	27.1896	41.4508	24.4415	37.2840	29.4125	44.8267	25.1164	38.3054	28.8707	44.0049	21.2321	32.4433	30.8379	46.9939
CS	0	0	0	100.389	104.2477	93.0893	96.6674	106.8175	110.9234	95.2555	98.9169	104.9569	108.9912	86.8363	90.1741	110.9009	115.1637
			0.2	46.1195	47.8922	42.6130	44.2510	49.1667	51.0566	43.6541	45.3321	48.2796	50.1353	39.5531	41.0734	51.0678	53.0307
		2	0	54.8714	82.8522	50.8815	76.8276	58.3852	88.1577	52.0655	78.6154	57.3682	86.6221	47.4637	71.6670	60.6171	91.5278
			0.2	25.2084	38.063	23.2918	35.1690	26.8740	40.5779	23.8608	36.0282	26.3890	39.8457	21.6192	32.6436	27.9130	42.1468
	200	0	0	99.0642	103.7257	91.6431	95.9762	105.5485	110.5016	93.8570	98.2870	103.6689	108.5377	85.1875	89.2444	109.6457	114.7851
			0.2	44.3306	46.5310	40.3507	42.4157	47.6185	49.9461	41.5422	43.6469	46.6520	48.9431	36.2210	38.2148	49.6051	52.0155
		2	0	53.9695	82.1957	49.8692	75.9578	57.5389	87.6275	51.0954	77.8229	56.5035	86.0520	46.2767	70.4955	59.7887	91.0520
			0.2	23.8350	36.3388	21.5203	32.8313	25.7025	39.1737	22.2156	33.8845	25.1512	38.3371	18.8920	28.8779	26.8142	40.8634
SS	0	0	0	85.5385	88.8264	78.7171	81.7428	91.6377	95.1601	81.0686	84.1847	89.5422	92.9840	73.3391	76.1581	95.2488	98.9099
			0.2	40.0876	41.6285	36.8675	38.2846	42.9313	44.5815	37.9794	39.4393	41.9513	43.5639	34.2827	35.6005	44.5865	46.3003
		2	0	46.7543	70.5959	43.0258	64.9661	50.0881	75.6297	44.3112	66.9069	48.9427	73.9002	40.0863	60.5276	52.0618	78.6099

(continued)

Table 6 (continued)

Thickness variation→			Uniform		Linear				Parabolic				Quadratic			
	(α, β)→		(0, 0)		(−0.3, 0)		(0.3, 0)		(0, −0.3)		(0, 0.3)		(−0.3, −0.3)		(0.3, 0.3)	
	$\Delta T\downarrow$	$p\downarrow$	$\gamma\downarrow$	TD	TI	TD	TI	TD	TI	TD	TI	TD	TI	TD	TI	TD	TI
			0.2	21.9114	33.0848	20.1513	30.4271	23.4657	35.4317	20.7591	31.3449	22.9301	34.6229	18.7385	28.2939	24.3705	36.7978
	200	0	0	84.2686	88.2474	77.3158	80.9886	90.4310	94.6860	79.7200	83.4980	88.3138	92.4739	71.7312	75.1682	94.0599	98.4794
			0.2	38.4619	40.3781	34.8260	36.6170	41.5154	43.5510	36.0664	37.9014	40.4701	42.4642	31.3056	33.0388	43.2472	45.3549
		2	0	45.8718	69.8673	42.0255	64.0165	49.2668	75.0335	43.3575	66.0424	48.1004	73.2586	38.9092	59.2792	51.2610	78.0686
			0.2	20.6598	31.5002	18.5492	28.3017	22.3907	34.1282	19.2654	29.3875	21.7998	33.2310	16.2990	24.9181	23.3604	35.6021
CF	0	0	0	59.4135	61.6972	56.3079	58.4723	62.3288	64.7246	58.0847	60.3174	60.9489	63.2916	55.3980	57.5269	63.8010	66.2531
			0.2	35.4504	36.8130	33.6370	34.9300	37.2015	38.6314	34.8226	36.1613	36.3370	37.7337	33.7283	35.0247	38.0752	39.5384
		2	0	32.4747	49.0346	30.7772	46.4715	34.0682	51.4406	31.7484	47.9379	33.3139	50.3017	30.2799	45.7203	34.8729	52.6556
			0.2	19.3768	29.2575	18.3856	27.7609	20.3339	30.7027	19.0336	28.7394	19.8614	29.9892	18.4355	27.8364	20.8115	31.4238
	200	0	0	58.5011	61.2662	55.2931	57.9210	61.4735	64.3691	57.1112	59.8184	60.0732	62.9067	54.2347	56.8278	62.9664	65.9283
			0.2	34.0086	35.7031	31.8172	33.4489	35.9596	37.7242	33.1300	34.8089	35.0326	36.7608	31.2633	32.9356	36.9106	38.7110
		2	0	31.8372	48.4923	30.0515	45.7771	33.4817	50.9938	31.0592	47.3098	32.7086	49.8176	29.4324	44.8389	34.3055	52.2468
			0.2	18.2666	27.8515	16.9599	25.8750	19.3902	29.5555	17.7157	27.0211	18.8650	28.7578	16.4656	25.1458	19.9320	30.3774

Table 7 Order of frequency parameter in first mode in terms of thickness variations

	Thickness decreasing from $X = 0$ to 1	Thickness increasing from $X = 0$ to 1
CC	$\Omega_{\text{Uniform}} > \Omega_{\text{Linear}} > \Omega_{\text{Parabolic}} > \Omega_{\text{Quadratic}}$	$\Omega_{\text{Uniform}} < \Omega_{\text{Linear}} < \Omega_{\text{Parabolic}} < \Omega_{\text{Quadratic}}$
CS	$\Omega_{\text{Uniform}} > \Omega_{\text{Parabolic}} > \Omega_{\text{Linear}} > \Omega_{\text{Quadratic}}$	$\Omega_{\text{Uniform}} < \Omega_{\text{Parabolic}} < \Omega_{\text{Linear}} < \Omega_{\text{Quadratic}}$
SS	$\Omega_{\text{Uniform}} > \Omega_{\text{Parabolic}} > \Omega_{\text{Linear}} > \Omega_{\text{Quadratic}}$	$\Omega_{\text{Uniform}} < \Omega_{\text{Parabolic}} < \Omega_{\text{Linear}} < \Omega_{\text{Quadratic}}$

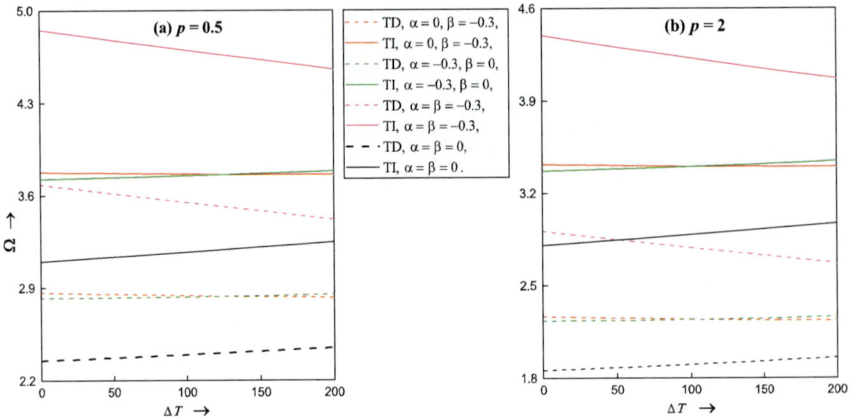

Fig. 4 Frequency parameter as a function of temperature difference ΔT for $\gamma = 0.1$ when the beam becomes thinner towards the right edge

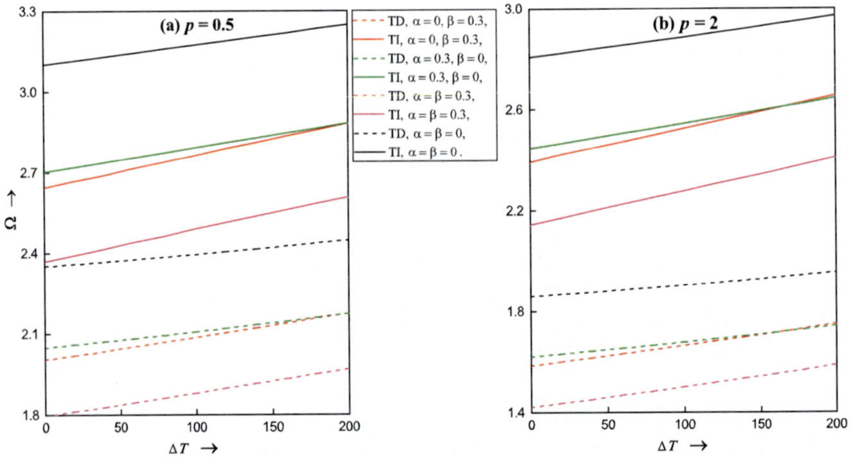

Fig. 5 Frequency parameter as a function of temperature difference ΔT for $\gamma = 0.1$ when the beam becomes thicker towards the right edge

is observed that the value of frequency parameter increases with the increase in the values of temperature difference for uniform and linear thickness variations, while it decreases for parabolic and quadratic thickness variations. This effect is more pronounced for temperature-dependent materials. Similarly, Fig. 5 represents the curves of frequency parameter for thickness increasing from left to right. It can be seen that the value of frequency parameter increases with the increase in temperature difference.

8 Conclusions

The vibrational behaviour of functionally graded nonuniform nanobeam under thermal environment is studied on the basis of Eringen's nonlocal elasticity theory. The effect of size dependency, ceramic/metallic constituents and surrounded thermal field, etc., is discussed for CC, CS, SS and CF boundary conditions and concluded that the advanced version of the differential quadrature method is giving accurate results with higher rate of convergence as compared to generalized differential quadrature method and generalized differential quadrature rule. Nanobeam with free boundary conditions shows paradoxical behaviour. However, for other boundary conditions, the increase in the values of temperature difference, volume fraction index and nonlocal parameter leads to lower stiffness of the beam, and so the frequency parameter decreases. Based on the discussion, one can deduce that the size effect plays a vital role in the vibrations of nanoscale beams-type structural elements.

9 Future Scope

Structural components of varying thicknesses are highly favoured these days due to their structural tailoring capabilities, damage tolerance and potential for creating significant weight savings in various engineering applications. Beam-type constructions of varying thickness are often encountered in the aerospace industry, missiles, spacecraft structures, satellites and transportation vehicles as well as in other fields of modern technology. In the past few decades, such structural elements are being used in the primary structures of an aircraft, such as the skin of the wing, the vertical fin torque box, aileron, spoiler, fuselage belly fairing and rotor blades. To use FGMs for the construction of beams efficiently, a good knowledge of the static and dynamic behaviour of such structures is essential with a fair amount of accuracy. Keeping the above in view, the following problems have their practical importance and should be discussed explicitly:

- The analysis of thick nonuniform functionally graded nanobeams in different technological situations.

- To study the composition of functionally graded materials with porous and piezoelectric materials.
- Impact of extremely harsh conditions on the performance of the above structural components.

Acknowledgements The financial support provided by the Indian Institute of Technology Delhi, India is gratefully acknowledged by the Dr. Rahul Saini, to carrying this work at IIT Delhi, India.

References

A.H. Akbarzadeh, A. Abedini, Z.T. Chen, Effect of micromechanical models on structural responses of functionally graded plates. Compos. Struct. **119**, 598–609 (2015). https://doi.org/10.1016/j.compstruct.2014.09.031

M. Aydogdu, A general nonlocal beam theory: its application to nanobeam bending, buckling and vibration. Phys. E Low-Dimension. Syst. Nanostruct. **41**, 1651–1655 (2009). https://doi.org/10.1016/j.physe.2009.05.014

O.A. Bauchau, J.I. Craig, *Structural Analysis* (2009)

D. Bernoulli, *Letters to Euler, Basel, 1735. Referencia extraida de Vibration of shells and plates* (Werner Soedel)

J. Biot, *Traite de Physique Experimentale et Mathematique* (Deterville, Paris, 1816)

V. Birman, L.W. Byrd, Modeling and analysis of functionally graded materials and structures. Appl. Mech. Rev. **60**, 195–216 (2007). https://doi.org/10.1115/1.2777164

A. Cauchy, *Exercices de mathematiques* (Paris, 1827)

S.-K. Chan, Y. Fang, M. Grimsditch, Z. Li, M.V. Nevitt, W.M. Robertson, E.S. Zouboulis, Temperature dependence of the elastic moduli of monoclinic zirconia. J. Am. Ceram. Soc. **74**, 1742–1744 (1991). https://doi.org/10.1111/j.1151-2916.1991.tb07177.x

E. Chladni, *Entdeckungen über die Theorie des Klanges* (Weidmann und Reich, Leipzig, 1787)

C. Coulomb, *Recherches theoriques et experimentales sur la force torsion et sur l'elasticite des fils de metal* (Paris, 1784)

W.H. Cubberly, *Metals Handbook*, 9th edn. (SME, 1989)

F. Ebrahimi, M. Ghadiri, E. Salari, S.A.H. Hoseini, G.R. Shaghaghi, Application of the differential transformation method for nonlocal vibration analysis of functionally graded nanobeams. J. Mech. Sci. Technol. **29**, 1207–1215 (2015). https://doi.org/10.1007/s12206-015-0234-7

M.A. Eltaher, S.A. Emam, F.F. Mahmoud, Free vibration analysis of functionally graded size-dependent nanobeams. Appl. Math. Comput. **218**, 7406–7420 (2012). https://doi.org/10.1016/j.amc.2011.12.090

A.C. Eringen, On differential equations of nonlocal elasticity and solutions of screw dislocation and surface waves. J. Appl. Phys. **54**, 4703–4710 (1983). https://doi.org/10.1063/1.332803

A.C. Eringen, *Nonlocal Continuum Field Theories* (Springer, New York, 2002)

L. Euler, *Methodus inveniendi lineas curvas maximi minimive proprietate gaudentes* (Berlin, 1744)

A. Farajpour, M. Danesh, M. Mohammadi, Buckling analysis of variable thickness nanoplates using nonlocal continuum mechanics. Phys. E Low-Dimension. Syst. Nanostruct. **44**, 719–727 (2011). https://doi.org/10.1016/j.physe.2011.11.022

F.A. Fazzolari, Natural frequencies and critical temperatures of functionally graded sandwich plates subjected to uniform and non-uniform temperature distributions. Compos. Struct. **121**, 197–210 (2015). https://doi.org/10.1016/j.compstruct.2014.10.039

T. Fuchiyama, N. Noda, Analysis of thermal stress in a plate of functionally gradient material. JSAE Rev. **16**, 263–268 (1995). https://doi.org/10.1016/0389-4304(95)00013-W

M.M. Gasik, Micromechanical modelling of functionally graded materials. Comput. Mater. Sci. **13**, 42–55 (1998). https://doi.org/10.1016/s0927-0256(98)00044-5

R. Kienzler, H. Altenbach, I. Ott, *Theories of Plates and Shells: Critical Review and New Applications* (Springer-Verlag, Berlin, Heidelberg, 2004)

J.H. Kim, G.H. Paulino, An accurate scheme for mixed-mode fracture analysis of functionally graded materials using the interaction integral and micromechanics models. Int. J. Numer. Methods Eng. **58**, 1457–1497 (2003). https://doi.org/10.1002/nme.819

M. Koizumi, The concept of FGM. Ceram. Trans. Func. Grad. Mater. **34**, 3–10 (1993)

E. Kröner, Elasticity theory of materials with long range cohesive forces. Int. J. Solids Struct. **3**, 731–742 (1967). https://doi.org/10.1016/0020-7683(67)90049-2

R. Lal, R. Saini, Vibration analysis of functionally graded circular plates of variable thickness under thermal environment by generalized differential quadrature method. J. Vib. Control. **26**, 1–15 (2019a). https://doi.org/10.1177/1077546319876389

R. Lal, R. Saini, On radially symmetric vibrations of functionally graded non-uniform circular plate including non-linear temperature rise. Eur. J. Mech. A/Solids. **77**, 103796 (2019b). https://doi.org/10.1016/j.euromechsol.2019.103796

R. Lal, R. Saini, On the high-temperature free vibration analysis of elastically supported functionally graded material plates under mechanical in-plane force via GDQR. J. Dyn. Syst. Meas. Control. **141**, 101003 (2019c). https://doi.org/10.1115/1.4043489

R. Lal, R. Saini, Vibration analysis of FGM circular plates under non-linear temperature variation using generalized differential quadrature rule. Appl. Acoust. **158**, 107027 (2020). https://doi.org/10.1016/j.apacoust.2019.107027

G. Lamè, *Lecons sur la theorie mathematique de l'elasticite des corps solides* (Paris, 1852)

L. Librescu, S.Y. Oh, O. Song, Thin-walled beams made of functionally graded materials and operating in a high temperature environment: vibration and stability. J. Therm. Stress. **28**, 649–712 (2005). https://doi.org/10.1080/01495730590934038

L.S. Liu, Q.J. Zhang, P.C. Zhai, The optimization design of metal/ceramic FGM armor with neural net and conjugate gradient method. Mater. Sci. Forum. **423–425**, 791–796 (2003). https://doi.org/10.4028/www.scientific.net/msf.423-425.791

A.E.H. Love, *A Treatise on the Mathematical Theory of Elasticity* (Dover Publications, New York, 1944)

P. Malekzadeh, A. Alibeygi Beni, Free vibration of functionally graded arbitrary straight-sided quadrilateral plates in thermal environment. Compos. Struct. **92**, 2758–2767 (2010). https://doi.org/10.1016/j.compstruct.2010.04.011

Q. Mao, S. Pietrzko, *Control of Noise and Structural Vibration: A MATLAB®-Based Approach* (Springer Science & Business Media, 2013)

A.J. Markworth, K.S. Ramesh, W.P. Parks, Modelling studies applied to functionally graded materials. J. Mater. Sci. **30**, 2183–2193 (1995). https://doi.org/10.1007/BF01184560

E. Müller, Č Drašar, J. Schilz, W.A. Kaysser, Functionally graded materials for sensor and energy applications. Mater. Sci. Eng. A **362**, 17–39 (2003). https://doi.org/10.1016/S0921-5093(03)00581-1

R.G. Munro, Evaluated material properties for a sintered α-alumina. J. Am. Ceram. Soc. **80**, 1919–1928 (1997). https://doi.org/10.1111/j.1151-2916.1997.tb03074.x

R. Nazemnezhad, S. Hosseini-Hashemi, Nonlocal nonlinear free vibration of functionally graded nanobeams. Compos. Struct. **110**, 192–199 (2014). https://doi.org/10.1016/j.compstruct.2013.12.006

M.Z. Nejad, A. Hadi, Eringen's non-local elasticity theory for bending analysis of bi-directional functionally graded Euler-Bernoulli nano-beams. Int. J. Eng. Sci. **106**, 1–9 (2016a). https://doi.org/10.1016/j.ijengsci.2016.05.005

M.Z. Nejad, A. Hadi, Non-local analysis of free vibration of bi-directional functionally graded Euler-Bernoulli nano-beams. Int. J. Eng. Sci. **105**, 1–11 (2016b). https://doi.org/10.1016/j.ijengsci.2016.04.011

M.Z. Nejad, A. Hadi, A. Rastgoo, Buckling analysis of arbitrary two-directional functionally graded Euler-Bernoulli nano-beams based on nonlocal elasticity theory. Int. J. Eng. Sci. **103**, 1–10 (2016). https://doi.org/10.1016/j.ijengsci.2016.03.001

N. Noda, Thermal stresses in functionally graded materials. J. Therm. Stress. **22**, 477–512 (1999). https://doi.org/10.1080/014957399280841

V. Panc, *Theories of Elastic Plates* (Noordhoff International Publishing, Leyden, The Netherlands, 1975)

B. Paszkiewicz, R. Paszkiewicz, M. Wosko, D. Radziewicz, B. Ściana, A. Szyszka, W. Macherzynski, M. Tłaczała, Functionally graded semiconductor layers for devices application. Vacuum **82**, 389–394 (2007). https://doi.org/10.1016/j.vacuum.2007.06.008

G.H. Paulino, Z.H. Jin, R.H. Dodds, 2.13—Failure of functionally graded materials. Compr. Struct. Integr. **2**, 607–644 (2007). https://doi.org/10.1016/B0-08-043749-4/02101-7

M. Petyt, *Introduction to Finite Element Vibration Analysis* (Cambridge University Press, 2010)

S. Poisson, *Sur l'equilibre et le mouvement des corps elastiques* (Paris, 1829)

S.C. Pradhan, Buckling of single layer graphene sheet based on nonlocal elasticity and higher order shear deformation theory. Phys. Lett. Sect. A Gen. At. Solid State Phys. **373**, 4182–4188 (2009). https://doi.org/10.1016/j.physleta.2009.09.021

J. Qiu, J. Tani, T. Ueno, T. Morita, H. Takahashi, H. Du, Fabrication and high durability of functionally graded piezoelectric bending actuators. Smart Mater. Struct. **12**, 215–221 (2003)

O. Rahmani, O. Pedram, Analysis and modeling the size effect on vibration of functionally graded nanobeams based on nonlocal Timoshenko beam theory. Int. J. Eng. Sci. **77**, 55–70 (2014). https://doi.org/10.1016/j.ijengsci.2013.12.003

S. Rao, *Mechanical Vibrations* (Pearson Education (Singapore) Pvt. Ltd., Indian Branch, Delhi, India, 2004)

J.N. Reddy, Nonlocal theories for bending, buckling and vibration of beams. Int. J. Eng. Sci. **45**, 288–307 (2007). https://doi.org/10.1016/j.ijengsci.2007.04.004

J.N. Reddy, *Theory and Analysis of Elastic Plates and Shells* (2008).https://doi.org/10.1002/zamm.200890020

T. Reiter, G.J. Dvorak, V. Tvergaard, Micromechanical models for graded composite materials. J. Mech. Phys. Solids. **45**, 1281–1302 (1997). https://doi.org/10.1016/S0022-5096(97)00007-0

R. Saini, R. Lal, Axisymmetric vibrations of temperature-dependent functionally graded moderately thick circular plates with two-dimensional material and temperature distribution. Eng. Comput. (2020). https://doi.org/10.1007/s00366-020-01056-1

B. Saint-Venant, Memoir sur les vibrations tournantes des verges elastiques. Comptes Rendus. **28** (1849)

I.H. Shames, C.L. Dym, *Energy and Finite Element Methods in Structural Mechanics* (Hemisphere Publishing Corporation, 1985)

H.S. Shen, Functionally graded materials: nonlinear analysis of plates and shells (2016)

C. Shu, Differential quadrature and its applications in engineering, Springer, 2000.

M. Şimşek, Bi-directional functionally graded materials (BDFGMs) for free and forced vibration of Timoshenko beams with various boundary conditions. Compos. Struct. **133**, 968–978 (2015). https://doi.org/10.1016/j.compstruct.2015.08.021

W. Soedel, *Vibrations of Shells and Plates*, 3rd edn. (CRC Press, Boca Raton, 2004)

A.H. Sofiyev, Review of research on the vibration and buckling of the FGM conical shells. Compos. Struct. **211**, 301–317 (2019). https://doi.org/10.1016/j.compstruct.2018.12.047

K. Swaminathan, D.M. Sangeetha, Thermal analysis of FGM plates—a critical review of various modeling techniques and solution methods. Compos. Struct. **160**, 43–60 (2017). https://doi.org/10.1016/j.compstruct.2016.10.047

R. Szilard, *Theory and Analysis of Plates: Classical and Numerical Methods* (Prentice Hall, Englewood Cliffs, New Jersey, 1974)

Y. Tanigawa, Some basic thermoelastic problems for nonhomogeneous structural materials. Appl. Mech. Rev. **48**, 287–300 (1995). https://doi.org/10.1115/1.3005103

S. Timoshenko, S.W. Krieger, *Theory of Plates and Shells*, 2nd edn. (McGraw-Hill Book Company, New York, 1984)

Y.S. Touloukian, *Thermophysical Properties of Matter 1* (1973)

E. Ventsel, T. Krauthammer, *Thin Plates and Shells: Theory: Analysis and Aplications* (CRC Press, 2001)

X. Wang, Differential quadrature and differential quadrature based element. Methods (2015). https://doi.org/10.1016/c2014-0-03612-x

C.M. Wang, J.N. Reddy, K.H. Lee, *Shear Deformable Beams and Plates: Relationship with Classical Solutions* (Elsevier, 2000)

Z. H. Wang, X. H. Wang, G. D. Xu, S. Cheng, T. Zeng, Free vibration of two-directional functionally graded beams. Compos. Struct. **135**, 191–198 (2016). https://doi.org/10.1016/j.compstruct.2015.09.013

F. Watari, A. Yokoyama, M. Omori, T. Hirai, H. Kondo, M. Uo, T. Kawasaki, Biocompatibility of materials and development to functionally graded implant for bio-medical application. Compos. Sci. Technol. **64**, 893–908 (2004). https://doi.org/10.1016/j.compscitech.2003.09.005

Y.Y. Yu, *Linear and Nonlinear Dynamical Modeling of Sandwiches, Laminated Composites, and Piezoelectric Layers* (Springer Science & Business Media, 1996)

J.R. Zuiker, Functionally graded materials: choice of micromechanics model and limitations in property variation. Compos. Eng. **5**, 807–819 (1995). https://doi.org/10.1016/0961-9526(95)00031-H

Current Prospective of Nanomaterials in Agriculture and Farming

Kamla Dhyani, Sobha, Maninder Meenu, Achintya N. Bezbaruah, Kamal K. Kar, and Pankaj Chamoli

Abstract Recently, nanotechnology has gained an intense attention in agriculture and quality farming system to meet the demand of sustainable agriculture. The unique properties of nanomaterials at nanoscale enables their employment for the design and development of diverse range of novel tools that supports sustainable agriculture. It is popular among the scientists due to its positive impact on agrifood sector by reducing the adverse impact of agripractices on environment, human health and improving food quality and productivity. In present chapter, application of various kinds of nanoparticles (NPs) in stress management of crops, as pesticides, herbicides, as nanobiosensors for disease detection, as seed growth promotes, for management of agricultural waste and shelf-life enhancement of agriproduce has been discussed in detail. Nanotechnology in agriculture significantly reduced the wastage of natural resources such as water, biofertilizers and also reduces the environmental pollution by reducing the application of harmful chemical fertilizers and pesticides. The application of nanomaterials found to be beneficial for sustainable agriculture. The recently available literature revealed the positive impact of nanotechnology application in different practices of agriculture such as crop nutrient management, stress resistance, insect and pest management, agriculture waste management, improving food security and productivity that, in turn, meet the food demands of global population.

K. Dhyani · Sobha
School of Agriculture Sciences, Shri Guru Ram Rai University, Dehradun, Uttarakhand 248001, India

M. Meenu
National Agri-Food Biotechnology Institute, Sector 81, Mohali, Punjab 140308, India

A. N. Bezbaruah
Nanoenvirology Research Group, Civil, Construction and Environmental Engineering, North Dakota State University, Fargo, ND 58105, USA

K. K. Kar
Advanced Nanoengineering Materials Laboratory, Materials Science Programme, Indian Institute of Technology Kanpur, Kanpur 208016, India

P. Chamoli (✉)
Department of Physics, School of Basic and Applied Sciences, Shri Guru Ram Rai University, Dehradun, Uttarakhand 248001, India
e-mail: pankajchamoli@sgrru.ac.in

J. K. Katiyar et al. (eds.), *Nanomaterials for Advanced Technologies*,
https://doi.org/10.1007/978-981-19-1384-6_9

Keywords Nanotechnology · Nanoparticles · Nutrient management · Nanobiosensors · Food packaging · Sustainable development

1 Introduction

Nanotechnology exhibits great potential in production and processing of agriproduce. Researchers are using biodegradable waste for green synthesis of nanoparticles (NPs). During green synthesis, the various plant secondary metabolites such as phenolic compounds, alkaloids, co-enzymes and terpenoids are reduced as NPs. The application of these NPs exhibits positive response in plant disease control, promoting plant growth, development and plant nutrient availability by the site-specific delivery system. For herbicide and pesticide application, encapsulated nanomaterials showed better penetration and allow slow release of herbicides and pesticides in the plant cell; therefore, nanotechnology provides an environment-friendly technique for herbicide and pesticide application in agriculture (Fig. 1) (Schils et al. 2018a). With advancement in the tool and techniques, the site selected delivery of pesticides and herbicides will further improve the agricultural practices. NPs loaded with fungicides, herbicides, fertilizers, nutrients and nucleic acid have a great potential to release at a specific part of the plant to achieve a site-specific control of plant diseases, abiotic stress management and nutrient management. With the advancement in biotechnological tools and techniques, nanomaterials have been extensively employed as disease diagnosis tools for advanced and precise diagnosis. The application of inorganic (micro or macro) nanomaterials during the growth cycle of crop increased the growth of plant by providing the appropriate nutrition to the plants

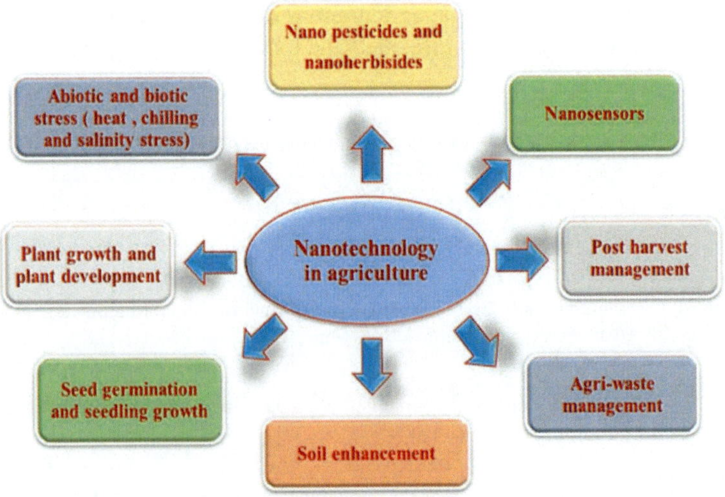

Fig. 1 Different aspects of nanotechnology in agriculture

(Schils et al. 2018a; Prasad et al. 2012). The metal-oxides NPs are proved to be effective to enhance the growth of plants in dry seasons by acting as a growth stimulator that, in turn, enhances the productivity and nutritional quality of the agricultural produce (Prasad et al. 2012; López-Vargas et al. 2018; Yang et al. 2006). In addition, in modern agricultural practices, nanomaterials are also reported to be useful as an effective antipest material (Bellesi et al. 2019). Nanotechnology deals with the use of NPs of dimensions 100 nm or less to meet the concept of quality and precision agriculture that enhanced the yield of the crop by efficiently utilizing resources in a precise manner by reducing the load of unnecessary chemicals on soil. (Auffan et al. 2009). Several materials like semiconductors, metal oxides, magnetic ceramics, synthetic and natural polymer lipids have been extensively explored for application of nanotechnology in agriculture. The biological material chitosan in nanomaterial as bionanocomposite is well known due to potent biopesticide and fungicide properties of chitosan that is helpful for seed priming treatment against fungal infection (Puoci et al. 2008). Furthermore, the plant response towards NPs varies from plant to plant, depending upon the uptake mechanism and their impact on growth and development. Seed germination and plant growth are the two major phenomena that are highly affected by the various concentration of NPs (Zheng et al. 2005).

In addition, researchers are emphasizing on green synthesis of NPs by employing various biological sources with an aim to reduce the adverse impact on environment (Bansal et al. 2014). The application of nanofertilizer, nanopesticides and nanoherbicides reported to significantly enhance the quality and quantity of agriproduce in the modern agriculture system. It is known that cereals are the most important staple food crop (Schils et al. 2018b); the foliar and soil application of NPs on cereal crops have positive effects on their productivity during their growth cycle that, in turn, increases the rate of plant productivity by offering micronutrient source (Cik et al. 2019).

1.1 Role of NPs in Seed Germination

Researchers have also revealed the positive impact of seed treatment with NPs that result in enhanced rate of seed germination and enhanced adaptation of seeds towards environmental stress (Adhikari et al. 2016). The seed treatment with NPs also reported to enhance seedling growth, vigour and viability in an experiment seed priming with Fe NPs which was found to be very effective in case of seedling growth parameters like root length, shoot length, pigments and antioxidant potential in Triploid water melon (*Citrullus lanatus*) (Kasote et al. 2019). Researchers have also revealed that the coating of seeds with silver (Ag) NPs resulted in enhanced water absorption compared to the control seeds (Adhikari et al. 2016). In another seed germination study, the seeds treated with NPs showed 73% more dry biomass and three times more vitamins compared to control seeds (Dehkourdi and Mosavi 2013). In addition, seed treatment with NPs also resulted in 90% increase drought resistance compared to control (Jaleel 2009). Furthermore, polysuccinimide NPs

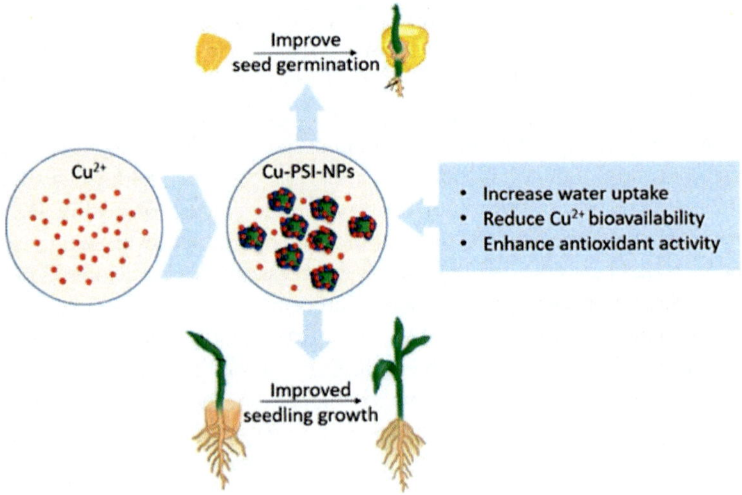

Fig. 2 Impact of polysuccinimide NPs (PSI-NPs) on corn (*Zea mays* L.) seed germination and seedling growth under Cu stress. Reproduced with permission of Xin et al. (2020)

(PSI-NPs) influenced the corn (*Zea mays L.*) seed germination and seedling growth under Cu stress (Fig. 2) in dose-dependent manner with an optimal rate of 200 mg L^{-1} (Xin et al. 2020) Thus, it can be assumed that nanoparticle treatment of seed or seedling results in significant increase in the quality, quantity and resistance of crop plants towards climate/environmental changes (Khodakovskaya 2009).

1.1.1 Plant Growth and Development

It is well known that the plant growth and development are significantly affected by the environmental regimes. The chemical treatment of plant propagating material could also promote or decrease the growth as well as the germination seed or seedling (Singh et al. 2015). Recently, several reports have been published related to the effect of nanomaterials on plants growth and germination and their advanced application in the agriculture field. The impact of various NPs such as titanium dioxide (TiO_2), copper (Cu), silicon (Si), gold (Au), quantum dots (QD), palladium (Pd), aluminium oxide (Al_2O_3), zinc oxide (ZnO), aluminium (Al) and cerium oxide (CeO_2) on germination of various plants namely, tomato, spinach, rice, lettuce, canola, radish, rye, grass, rape, corn, cucumber, cabbage and wheat was explored (Xin et al. 2020). These studies mentioned the positive impact of NPs on seed germination rate, photosynthetic rate, biomass and chlorophyll content of plants. In general, the seed germination rate was reported to be inversely correlated with the nanoparticle size. In addition, NPs mentioned to enhance the absorption level of inorganic nutrients, stimulate organic substance disintegrations and also improve the photosynthetic rate in plants

(Shojaei 2009). Furthermore, it is suggested to conduct extensive studies for deciphering the engineered nanomaterials, exploring their mechanistic application along with the agroecological toxicity (Khot 2012; Daniel and Astruc 2004; Kato 2011).

1.2 Role of Nanotechnology in Agriculture Innovation

Nanotechnology can play an important role to achieve the goal of sustainable agriculture and quality farming systems development. It is of immense importance to meet the food demands of increasing human and animal population. Thus, it is crucial to mindfully employ the available technologies for sustainable development. During the green revolution, fertilizers, pesticides and other agrochemicals have been used in excess with an aim to enhance the yield of crops. After decades of this unmindful practice, it was observed that excessive use of these agrochemicals leads to irreversible soil and environment pollution and reduction in soil microflora. To address this issue, application of nanotechnology in agriculture plays an important role by using significantly less amount of chemicals to achieve high protection against pests, herbs and high yield of agriproduce. However, it is also important to consider that excessive use of nanofertilizers, nanopesticides and nanoherbicides can also be toxic and harmful to the environment. Thus, further research is required to optimize the amount of nanomaterial to be used to achieve the goal of sustainable agriculture. Furthermore, nanotechnology can be helpful to improve the quality and production of agriproduce by employing nano-based sensors and monitoring devices in various agricultural practices. Thus, appropriate use of nanotechnology in agriculture may increase global food production and it will affect world agriculture positively (Kato 2011). Nanotechnology is an emerging field in the twenty-first century. Globally, researchers are exploring various means for the commercialization of nanoproducts. NPs have gained considerable attention compared to bulk counterparts owing to their unique properties. ZnO NPs mentioned to exhibit great potential to enhance crop yield due to their potent physical, optical and antimicrobial properties. The metal and metal-oxide NPs were also reported as an effective growth stimulator of crops in dry seasons resulting in an enhanced yield and nutritional quality of agricultural products by acting as an active antimicrobial agent. Nanomaterials are also employed as an essential part of different biotic and abiotic remediation strategies as NPs play a significant role in deciding the fate, mobility and toxicity of soil pollutants. When nanomaterials entered in the soil system, they may exhibit significant impact on soil quality as well as plant growth and development that has been conversed as their effect on nutrient release in soil, soil organic matter, soil biota as well as physiological and morphological responses of plants. In addition, the mechanisms involved in nanomaterial uptake and translocation within plants, as well as associated defence systems, are addressed in the following sections.

1.3 *Useful NPs in Agriculture*

NPs are also reported to play a significant role in plant growth, development and productivity. NPs' biological function is determined by their physicochemical qualities, application method and concentration. Previous studies have reported several NPs to be used in agri-innovation. It was also mentioned that different types of NPs exhibit different effect on plant materials such as increase in the seed germination, biomass or grain yield. Some of the NPs are toxic in nature; thus, it is recommended to use eco-friendly NPs synthesized by employing green methods and biodegradable and safe materials. Recently, researchers are also exploring different biological agents such as viruses, fungi, bacteria and plant extracts for NPs synthesis as these agents are green reducing agents and reduce the environmental impacts. Overall, Ag NPs, TiO_2 NPs and ZnO NPs are the most commonly employed NPs in agriculture for various purposes.

1.3.1 Carbon NPs

Carbon NPs have been used in different forms in several plant growth studies; C_{60} fullerene is the first stage carbonaceous NPs (Mukhopadhyay 2014). Carbon NPs of size 10 nm are the most advanced types of carbon nanomaterials. It is a nascent fluorescent molecule because of its unique trait of high photoluminescence that is proportional to its size. The cost, size, water solubility, transparency and biocompatibility of carbon NPs all play a role in their practical application. Based on its structure, two main types of nanotubes are available single-walled and multi-walled nanotubes. Carbon NPs have recently been added to the importance and extensive use of carbon materials. An extensive research in this field results in the development of novel exciting carbon nanomaterials that have attracted significant attention of researchers from various fields such as water filtration, hydropower, biochemical and agriculture production (Baker and Baker 2010). In literature, different types of methods have been reported for the synthesis of carbon NPs such as laser isolation, arc discharge, carbonization of carbohydrates, microwave-based pyrolysis and chemical vapour deposition (Mostofizadeh 2011; Chamoli et al. 2017; Singh et al. 2020; Wang 2011). Thus, by employing current breakthroughs in the field of nanotechnology, major improvements and enhancements in agricultural sustainability, disease management, crop protection, variety improvement and productivity can be realized. (Fig. 3) (Patel et al. 2019).

1.3.2 Metal-Oxide NPs

Ag NPs exhibit potent antimicrobial activity attributed to its high surface area (Wei 2013). Ag NPs have been extensively used against various disease-causing broad-spectrum human and plant pathogens (Cho et al. 2005; Morones et al. 2005; Tian

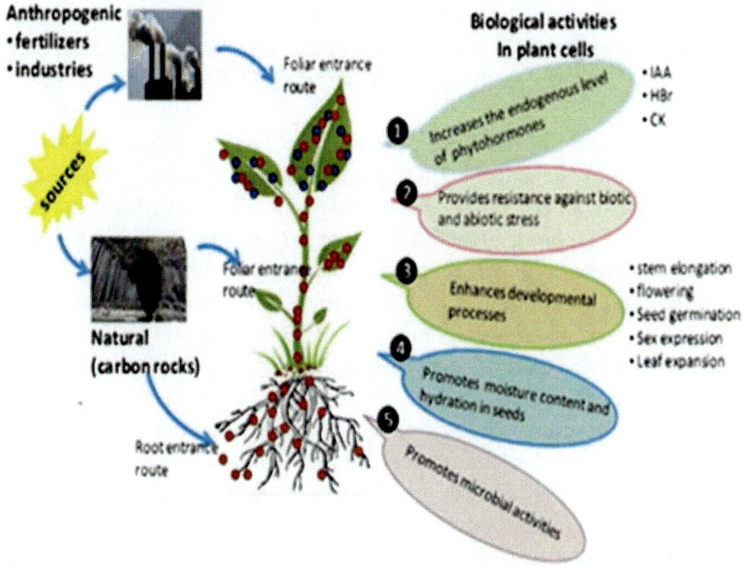

Fig. 3 Schematic representation of carbon nanotube sources (natural and anthropogenic), their uptake, accumulation and translocation, and their impact on growth and development of crop plants. CK, Cytokinin; HBr, homobrassinolide; IAA, indole-3-acetic acid. Reproduced with permission Patel et al. (2019)

et al. 2007; Chamoli et al. 2021a, b; Shukla et al. 2019; Ali et al. 2015). These particles have also been successfully employed for pest control of major food crops. Several methods for synthesis of Ag NPs have been documented in the literature, including chemical, physical and biological processes. Researchers have recently focused on environmentally safe, single-step methods for the synthesis of Ag NPs. Ag NPs have also been synthesized from a variety of sources, including plants, bacteria, and fungi. These Ag NPs are successfully employed for plant pathogen control in food crops. Researchers have also investigated the impact of Ag NPs (diameter = 20 nm) on fenugreek plant with positive results of pest control (Hojjat 2015). Researchers also found that seeds treated with Ag NPs at a concentration of 10 $\mu g\ ml^{-1}$ exhibit the highest seed germination, germination speed, root length and root fresh weight. These results clearly revealed that the positive impact of Ag NPs on seed germination (Hojjat 2015). Ag NPs also reported to exhibit antibacterial properties, against the rice pathogen (*Xanthomonas oryzae* pv. *oryzae* (Xoo) which cause bacterial leaf blight (BLB) disease, NP synthesized from susceptible rice variety was used as an antibacterial agent against phytopathogen and results found to be very effective in mitigating the bacterial growth and colony formation of Xoo; therefore by this experiment, Ag NPs were found to be more powerful antibacterial agent (Namburi et al. 2021). Similarly, the Ag NPs coated with fructose reported to present antimicrobial activity against phytopathogens such as Erwinia amylovora, Clavibactermichiganensis, Ralstonia solanacearum, Xanthomonas campestris and

Dickeyasolani (Mishra et al. 2014). Zinc is also an important micronutrient for different agricultural crops. The deficiency of Zn significantly affects the agricultural production especially in the calcium carbonate-rich soil (Mishra et al. 2014). The soil of the Mediterranean region and arid region is mostly supplemented with calcium carbonate and this soil limits the zinc availability to plants (Thwala et al. 2013). ZnO NPs may be quite helpful to address this problem of zinc deficiency. However, exploration of harmful effect of ZnO NPs is crucial while addressing the problem of Zn deficiency of plants in calcium carbonate-rich soil (Elumalai et al. 2015). ZnO NPs increase the zinc dissolution and bioavailability of Zn to the plants in soil enrich with calcium carbonate. The diffusion of zinc from Zn fertilizer is an important part of zinc absorption by the roots of plants (Rajiv et al. 2015). ZnO NPs less than 100 nm in size exhibit better antimicrobial activity due to high interaction with bacteria as they exhibit high surface to volume ratio (Gangloff et al. 2006). ZnO NPs are also vital in the antioxidant defence system because they deactivate reactive oxygen species (ROS), which causes cell death. Thus, ZnO NPs are also reported to be helpful in the stress tolerance mechanism. The toxicity of Ag and ZnO NPs was suggested by the generation of reactive nitrogen species (RNS) and hydrogen peroxide (H_2O_2) when duckweed (*Spirodela punctuta*) was exposed to Ag and ZnO engineered NPs (Xia et al. 2006). In wheat, zinc NPs have been found to induce free radical production, resulting in increased malondialdehyde and reduced glutathione (Ryter et al. 2007) and chlorophyll concentration (Long et al. 2006; Lovric et al. 2005). To solve the zinc deficiency problem, ZnO and $ZnSO_4$ are mostly used zinc fertilizers but due to the non-availability of Zn to the plants, their use as Zn fertilizer is limited. ZnO NPs can address the above-mentioned problem by increasing the solubility of Zn and enhancing its availability to the plants. The enhanced solubility and bioavailability of Zn NPs attributed to its size in nanorange; thus, these NPs are more active as compared to Zn particles of millimetres in size (Xie et al. 2011).

1.4 Stress Management and Tolerance

1.4.1 Abiotic Stress Tolerance in Plants

Any adverse environmental condition such as temperature, moisture and salinity that affects the plants is known as abiotic stress. The abiotic stress of any kind can adversely affect the plants and limits their productivity. According to a report, global agricultural production should have been boosted to 70% to fulfil the fast-rising food demands of human population. Thus, increase in the different type of abiotic stresses and their adverse impact on crop yields triggers plant scientists to explore different means to control their impact on crop yield. (Lewinski et al. 2008). Under these abiotic stress conditions, plants have developed different mechanisms to combat these stress conditions. It is also important to note that the response towards these stress conditions may vary depending on the plants and their species. Therefore, screening or selection of stress tolerant genotype is a major concern for the plant scientists for sustainable agriculture. In this case, nanotechnology has opened new doors in the field of biotechnology and agriculture to deal with abiotic stress conditions such as salinity, alkaline soil condition, heavy metal stress, high- or low-temperature (FAO 2050; Meenu et al. 2016). Various physiological and molecular changes occur under abiotic stress conditions such as changes in gene expression, cell division and evolution of different energy pathways (Bromham et al. 2013). In such cases, NPs behave as stress signals and activate the defence mechanism of plants against abiotic stress conditions. The NPs also reported to act as antioxidative enzymes and scavenge the ROS (Manzer et al. 2015).

Drought

The depletion of water resources, desertification and salinity are the major challenges all over the globe in the field of agriculture, food production and food security. Among various abiotic stress, drought is the major problem as it limits the production of crops in arid regions (Rico et al. 2013). Recently, in case of *Crataegus spp.,* researchers have found that application of Si NPs at different levels of drought presented positive physiological and biochemical changes such as enhanced photosynthetic activity, MDA, high proline accumulation and more chlorophyll content (Wahid 2007). In addition, drought susceptible cultivars of Sorghum (*Sorghum bicolor* L.) presented an improved tolerance towards drought followed by the treatment with silicon NPs. Sorghum plants were also reported to exhibit improved root growth and photosynthetic rate. All these parameters revealed improvement in the drought tolerance of sorghum plants followed by treatment with silicon NPs (Rico et al. 2013).

Heat Stress

The exposure of food crops to high temperature more than the optimum is regarded as heat injury, temperature stress and heat stress. Heat stress severely affects the growth and development of plants when exposed for a long time (Rico et al. 2013). The exposure to heat stress or any other abiotic stress condition led to the production of ROS in that, in turn causes oxidative stress in plants that lead to ion leakage and lipid peroxidation. This will result in the degradation of some important proteins and reduction in photosynthesis rate and chlorophyll content (Wahid 2007). Se NPs had previously been shown to minimize the effects of heat stress by boosting chlorophyll content, plant growth and hydration ability at low concentrations. The expression of heat shock proteins (HSPs) is another important feature associated with heat stress. These HSPs work as molecular chaperones. The application of CNTs has also been reported to be linked with the HSP related genes (Ahmed et al. 2021). The excessive production of H_2O_2 and upregulation of HSP70 have also been reported to be associated with the application of CeO_2 NPs (Khodakovskaya et al. 2012). Furthermore, treatment with TiO_2 NPs was also shown to minimize the effects of heat stress by regulating stomatal opening (Zhao et al. 2012).

Chilling Stress

Chilling stress or low-temperature stress is the condition in mesophytes and xerophytes when they suddenly exposed to very low-temperature conditions, sometimes near freezing which causes serious damage in plant cells (Qi et al. 2013). The adverse effects of chilling stress include loss of permeability and ion leakage from the membrane, that, in turn lead to reduced germination, growth and overall development of plant (Hasanuzzaman et al. 2013). The tolerance to chilling stress varies significantly from variety to variety. The plants tolerant to chilling stress experience less damage while sensitive plants exhibit more damage followed by cold stress. The TiO_2 NPs were found to decrease the negative effects of chilling stress by minimizing plasma membrane damage, maintaining permeability and preventing ion leakage (Welti et al. 2002). The photosynthetic light reactions were also mentioned to be significantly influenced by the chilling stress. Chilling stress reported to exhibit several negative effects on plants such as reducing transpiration rate and CO_2 assimilation rate as chilling stress degrades RUBISCO and chlorophyll (Mohammadi et al. 2013). Furthermore, the nanoparticle treatment of plants leads to enhanced production of Rubisco enzyme in photosystem (Yordanova and Popova 2007). In addition, nanoparticle treatment also increases the light immersion ability of chloroplast (Gao et al. 2006a), as well as inhibits the production of ROS along with reducing the rate of ROS production. In addition, the plant treated with TiO_2 NPs presented high Rubisco activity (Ze et al. 2011), increased activity of the antioxidant enzyme (Xu et al. 2014), enhanced leaf pigments and increased tolerance to chilling stress. It was also mentioned that the upregulation of stress-related genes during low-temperature stress conditions increased the level of MDA reductase, glutathione reductase and dehydroascorbate reductase activities. These proteins/enzymes scavenge ROS that, in turn, reduces oxidative damage such as peroxidation of lipids, chlorophyll lose and generation of H_2O_2 and lead to enhanced tolerance to chilling stress (Hasanpour et al. 2015). Overall, nanoparticle exposure during chilling stress exhibits positive impact on growth, biochemical and physiological response towards the resistance (Haghighi et al. 2012a).

Salinity Stress

Application of NPs in salinity stress has been extensively explored. Among the NPs under investigation, SiO_2 NPs exhibit high potential to protect plant under salinity stress. A significant enhancement in the chlorophyll content, accumulation of proline and antioxidant enzyme activity was observed followed by the treatment with SiO_2 NPs that, in turn, will lead to enhanced abiotic stress tolerance in plants (Haghighi et al. 2012a). The treatment with Si NPs on lentil seed under salinity stress lead to a significant increase in the seed germination and seedling growth (Haghighi et al. 2012b). Furthermore, in the control condition, seed germination and seedling growth were lower than in the lentil seeds treated with SiO_2 nanoparticles under salinity stress. Thus, it can be assumed that SiO_2 NPs may increase the salinity tolerance in plants at seedling stage (Gholamreza et al. 2020; Sabaghnia and Janmohammad 2015). In maize, an increase in the fresh and dry weight of root and shoot was recorded followed by the application of SiO_2 NPs under salinity stress (Savvasd et al. 2009). Furthermore, Na toxicity is a major concern under salinity stress conditions, and it adversely affects crop growth and yield. In this regard, several studies reported positive impact of NPs treatment on plants along with increasing salinity tolerance in plants especially due to SiO_2 nanoparticle treatment. Thus, NPs treatment lead to overall improvement plant growth under stress/adverse conditions. Furthermore, salt stress raises Na toxicity, which reduced crop development and yield, and SiO2 NPs treatment was recommended to minimize Na ionic toxicity that result in improved crop growth and production and overall crop improvement under unfavourable conditions (Zulfikar & Asraf 2021).

Heavy Metal Stress

Heavy metal stress is another critical environmental issue being faced globally. Heavy metal stress suppressed the crop growth and yield by increasing the heavy metal toxicity by disturbing various physiological activities in plants (Gao et al. 2006b). Heavy metal stress interferes with the nutrient uptake process and also affects the regular antioxidant enzyme activity which exhibit positive effect on various activities associated with stress tolerance in crop plants (Rahimi et al. 2012). Heavy metal stress in soil and water increases the production of ROS that in turn increases oxidative damage in plants. The oxidative damage increases the stress condition plants by altering the structure of cells and degradation of various crucial proteins and enzymes. The stress conditions and oxidative damage decrease the nutrient uptake in plants that ultimately lead to nutrient deficiency and poor enzyme activity that, in turn, reduced the growth and development of plant (Capuana 2011). In response to heavy metal stress, plants have developed various defence mechanisms such as polyphosphates and metal chelators production which limit the excessive uptake of heavy metals and activate the antioxidant enzymes system that scavenges ROS production. However, application of synthesized NPs reported to reduce the load of heavy metal stress or toxicity (Rascio and NavariIzzo 2011; Gunjan et al. 2014; Tripathi et al. 2015). Previously, it was shown that using TiO_2 and hydroxyapatite nanoparticles reduced cadmium toxicity and increased photosynthetic rate and plant growth in *Brassica juncea* (Worms et al. 2012). It was also mentioned that supplementing growth media with Si NPs reduces chromium toxicity in peas (Gunjan et al. 2014). Furthermore, cowpea exposed to gold ion stress reduced Au31 to non-toxic gold NPs in the presence of phenolic compounds from germinating seeds (Singh and Lee 2016).

1.5 Nanotechnology in Food Industry

In addition, the introduction of nanotechnology to generate new generation packaging material reported to significantly enhance the freshness and quality of the fresh agriproduce. These novel packaging materials protect the packaged material from harmful rays, gases, chemicals and hindered the growth of pathogenic microorganisms. However, selection of base material packaging development and fast disintegration of packaging environment are two important factors to consider (Shabnam et al. 2014). In this context, polymer nanocomposites have emerged as an absolute substitute (Stewart et al. 2002) and its use as a packaging material which increase or extend the consumption of digestible and non-toxic degradable films. The application of nanocomposites results in edible and biodegradable films that protect the food against nutrient loss along with providing protection against cancerous materials (Stewart et al. 2002). Due to the availability of limited amount of natural polymers and weak mechanical strength of reported nanocomposite packaging material, their commercial utilization is limited. Thus, these natural biopolymer materials are usually combined with other man-made polymers with an aim to improve their mechanical strength and commercial utilization (Sinha Ray and Okamoto 2003). The packaging food material with bionanocomposites reduces the environmental load of packaging waste. These bionanocomposites are also maintain the freshness of food material for longer time and significantly enhance the self-life of packaged food. These nanocomposite packaging materials are associated with advantages such as low surface thickness, lucidity, easy flow, better surface properties and show properties of easy recyclability (Petersen et al. 1999). Furthermore, in the field of bionanocomposites as packaging material, the photo-catalysts mechanism by TiO_2 NPs under UV irradiation has gained significant attention due to its significant antimicrobial activity that in turn enhances the shelf life of packaged food. In addition, the catalyzing ability of nano-TiO_2 in presence of UV light can oxidize ethylene into water and CO_2 (Chen and Hu 2005). The development of edible and biodegradable films from natural resources promoted the exploration of innovative bio-based packaging materials that lead to shelf-life enhancement of food material along with reducing packaging waste (Hu and Fu 2003). Recently, Fuji apples packaged with nano-SiOx/chitosan presented better food processing qualities in comparison to non-degradable polymers (Tharanathan 2003). The development of advanced packaging materials with nanomaterials is able to meet the demand of advanced preservation techniques for storage of perishable food products, fruits and vegetables, and different beverages. With the addition of suitable NPs, normal biodegradable packaging material can be converted into durable and heat resistant packaging material. Furthermore, to ensure the food safety, nanobiosensors can be employed to detect the presence of unwanted biochemical reaction, generation of gases and presence of microorganisms (Tharanathan 2003). The functional ingredients of fruits, such as vitamins, antioxidants are the fundamental components, but they are rarely employed in their purest form; functional ingredients are commonly used as a part of a delivery system rather than the natural ingredients of fruits. Several functions can be performed by employing delivery system; however, the main goal is to deliver that functional element to its specific location. However, protecting of an ingredient from biotic and abiotic degradation is an important function of a delivery system. In addition, delivery system also responsible for the release of functional ingredient under specific environmental situations. Nanodispersions and nanocapsules are specific delivery systems employed for drug delivery of functional ingredients to specific sites. Nanobiosensors are used for labelling products with biodegradable sensors. In food industry, change the food colour with the change in pH of food products due to degradation, spoilage followed by microbial activity was determined by employing biosensors. Biological molecules such as sugars or proteins are frequently employed in the food industry as biosensors to detect pathogens and contaminants (Whistler and Daniel 1990). Nanotechnology may also be beneficial in coating or priming of seeds and grains with suitable materials that protect them against biotic and abiotic stress by various environmental factors. It is also used to design several food materials with potent antioxidants properties and different flavours. The primary goal of coating with nanoparticles is to improve the quality and functioning of ingredients by lowering their concentrations or dilutions as well as to improve the product by lowering the presence of chemicals by infusing innovative substances into foods (Charych et al. 1996) that can be achieved with greater exploration of delivery and controlled-release systems for biopharmaceuticals (Haruyama 2003).

1.6 Nanotechnology in Insect and Pest Management

Regarding the implementation of nanotechnology in insect and pest management, the major focus is the use of NPs for plant protection and nutrition in the form of nanopesticides or fertilizers. The application of pesticides and fungicides plays a crucial role in advanced agricultural practices but the area of food processing and packaging the application of nanopesticides and nanofertilizers has received less attention. Due to the direct application of nanochemicals in nature, they may be degraded or diffused in the environment. And this may be a critical condition for diffused nanoagrochemicals (Lawrence and Rees 2000). For wide applications of nanotechnology, several new products are being explored. Recently, nano-based agrochemicals were applied in agricultural pest management and in other fields of food processing and packaging with an aim to provide plant protection, nutrition management, development of eco-friendly, renewable energy resources, management of biomass, biocomposites and agrochemical industries (Sadowski 2010). Nanoformulations of ZnO, Ag, Cu, SiO_2, exhibit a wide range of protection against pest and insects, water stress and act as sustainable substitute for conventional pesticides to control the quality of soil and environmental pollution compared to the other traditional methods (Sadowski 2010). Zinc is essential for plant growth and development, yet it is only found in trace amounts in soil. As a result, Zn metal is a key target for the production of Zn NPs for pest control and nutrient delivery to plants. These nanoparticles are both cost-effective and safe, with strong antibacterial and antifungal activities. Thus, these NPs can be efficiently employed in agricultural fields to post-harvesting management practices. In addition, Ag NPs also exhibit antibacterial, microbial, fungal, larvicidal, pesticides and antiviral activity and efficiently employed in the field of agriculture, plant health management and pest control during pre- and post-harvest practices of food and grains (Chhipa 2017; Chopra et al. 1994; Gao et al. 2014; Iravani 2011). The nano-based formulations with ZnO NPs against the microorganisms show stability and slow kinetics, which is used for production in large amounts of antifungal reagents. The antifungal activity of bioformulation of nanomaterials may be the best alternative against harmful insects-pest control systems. In several experiments, the researchers were found that when an aqueous solution of bio-NPs was applied on several insects and pests, they found results like in the case of larvicidal activity of *H. armegera*. The aqueous extract from *Ecliptaprostrata* is useful to control mosquito; aqueous extracts of *E. prostrate* is used to control *S. oryzae*. Aloin from Alovera on formed Ag NPs is very effective, which is an eco-friendly approach to control the attacks of various insects and pests (Rajakumar and Rahuman 2011; Kantrao et al. 2017; Logaranjan et al. 2016; Devi et al. 2014). The use of nanotechnology in food packaging shows high efficiency in terms of biodegradation, a better solubility of nutrients and their slow release in the soil maintain soil fertility that, in turn, lead to quality improvement of crops. As a result, future researchers can investigate the use of nanotechnology in the field of plant protection and management in order to solve the fundamental issues associated with chemical fertilizers.

The properties of biopesticides such as thermal stability, stiffness, permeability and biodegradability are more advantageous in comparison with chemical fertilizers. Pests may experience indigestion followed by ingestion of nano-based material, which disrupt their water protection barrier, resulting in desiccation and death. Due to the presence of nanosize, nano-based formulations with nanocarriers have efficiently decreased the population of insects and pests. Bioactive compounds initiate the release of secondary metabolites from plants which can act on microorganisms present in its periphery (Kamaraj et al. 2012). Several studies have been carried out to determine the toxic effect of NPs on bacteria, fungi and pathogens, as well as a few experiments on insects and pests such as *Amsacta moorei* Butle, *Brachytrypesportentosus* Licht, *Episomus lacerta* F, *Gryllulus Domesticus* Linn, Chrotogonus sp. and *Helicoverpa armigera* (Nuruzzaman et al. 2016). The identification of plant species harbouring secondary metabolites with insect repellent properties as expressed by NPs such as Ag, ZnO, TiO_2 has recently been a hot topic. For future application and human welfare, the impact of nano-based biologically produced NPs on environment and living beings should be explained.

1.7 Agricultural Waste Management

In the agriculture sector, there is the production of large amounts of biowaste in each step, from planting to processing to till food vegetable and fruits storage. Agriculture biowaste management is a huge problem due to the shortage of skilled labour, proper mechanization and availability of adequate infrastructure. Agricultural biowaste processing has various limitations ;due to that, only 2% of the whole world biowaste is used, and all the other useable biodegradable material is degraded or decomposed by microbes; due to that, a large amount of biowaste is lost in the form of crop and energy loss. If this biodegradable waste product is properly managed and used, then we have a huge source of renewable energy. This is easily created and used by the whole world. By the use of nanotechnology or nanobioengineering, the enzyme extraction efficiency and energy production could be increased, and agricultural waste is easily utilized as a major source of renewable bioenergy which is very helpful in the conservation of biological resources and sustainable agriwaste management, nature and natural products. In nanotechnology by using metalloid enzymes, the biofuel production capacity was increased from agricultural wastes like rice husk, sugarcane waste, vegetable oils, shells of coconut, cotton stalk, groundnut covering corncobs, cotton and animal fats (Shiva et al. 2020; Shrivastava and Dash 2012; Sarkar and Praveen 2016). Nanotechnology provides natural replacement of harmful chemicals and also induce degradation of pesticides and herbicides. These hazardous pollutants can be degraded and converted to harmless compounds using NPs as reactive agents (Bharati and Suresh 2017). It is well recognized that all chemicals mixed with wastewater have a dangerous effect on the environment, and it is critical to remove these waste products in a systematic manner (Ditta 2012; Babula and Farming 2009). For the treatment of wastewater, several strategies have been developed, including nanotechnology. One of the most important nano-based wastewater treatments is photocatalysis. Purification, filtration and decomposition as well as the elimination of pathogens such as bacteria, viruses and other harmful agents have all been accomplished using photocatalysis. Photocatalysis is a catalytic reaction that takes place in presence of light. Several NPs have been used as catalysts, including metal oxides and sulphides such as TiO2, ZnO, SnO2 and ZnS (Mulligan et al. 2001; Ko 2009).

1.8 Nanobiosensors: New Tool for Detecting and Diagnosing Crop Diseases

All organisms can sense the various environmental changes. Disease detection and diagnosis is an important prospect in plant protection and the combination of biology and nanotechnology proved to be helpful in this aspect. Before controlling the disease, its detection is very important. Nowadays, NPs may be for disease detection, these compounds that could sense the presence of the pathogen. Nanobiosensors are every minute reproducible, solid and less toxic. They are helpful to maximize sustainable agriculture. With the help of signal receptors, biosenseros receives signals from the environment. Nanobiosensors have three components probe, bioreceptor and transducer. In comparison to the standard ELISA method, nano-based bioformulations with biosensors have a higher sensitivity for pathogen detection and disease diagnosis (Patel et al. 2021; Feigl et al. 2010). Nanosensors are used in plant pathology to detect disease-causing agents, contaminants in the environment and the presence of nutrients in soil. (Elmer and White 2018). Insect attacks in the crop plant can also be detected using nanosensors based on the detection of chemical substances generated by insects. (Brock et al. 2011). According to a study, QD nanosensors were also constructed to detect the presence of lettuce tobacco and cowpea mosaic virus, as well as beet necrotic yellow vein virus (Chartuprayoon et al. 2013; Lin et al. 2014; Safarpour et al. 2012). Several portable nanodevices have also been developed to detect environmental contaminants, insects, pathogens and diseases (Sharon et al. 2010). In wheat, gold-based immunological sensors were employed to detect kernel bunt disease (Singh 2020). Plants accumulate various stress-related chemicals when they are under stress. Previously, by sensing salicylic acid concentrations in soil, a gold electrode nanosensor and copper NPs detected a plant pathogenic fungus (Shang et al. 2019).

2 Conclusions

Nanotechnology have various application in agriculture as discussed in the chapter, NPs have the potential to change the scenario of global food security and agriculture problems. Nanotechnology has showed considerable promise in agricultural applications, as it has the potential to improve people's lives and the global economy. The focus of this chapter is on providing basic understanding regarding the optimal use of nanotechnology and different NPs for the improvement of crop productivity and sustainable agriculture. As advancement in nanotechnology going on, there are various tools and techniques available for a variety of agricultural nanotechnology applications, such as use of NPs in DNA sequencing, nanobarcodes, nanosensors, nanocatalysts, nanofertilizers and nanopesticides. Many studies have been undertaken to investigate the impact of carbon NPs on plant growth and development activities; nevertheless, different plant species respond differently to different nanoparticles, and the reason behind these variations are unknown. As a result of various studies, NPs are most popular due to their versatile properties, for water and wastewater purification. Ferrite-based adsorbents have low toxicity, high chemical stability and are economical to use as they can be easily separated from the purified liquid. Ferrites have shown great promise and are potential candidates for application for water and wastewater. Excellent magnetic properties make the use of ferrites in water systems attractive as they can easily be recovered at the end of the treatment train using a conventional magnetic field. While they can easily use in water treatment systems, their use in wastewater systems would need additional work given the complexity of wastewater. As a result, ferrites adsorbents are a top choice due to their versatile properties, reasonability and magnetic separation capability for water and wastewater purification.

References

T. Adhikari, S. Kundu, A.S. Rao, Zinc delivery toplants through seed coating with nano-zinc oxide particles. J. Plant Nutr. **39**(1), 136–146 (2016)

T. Ahmed, M. Noman, N. Manzoor, M. Shahid, L. Ali, G. Wang, A. Hashem, A.B. Al-Arjani, F. Allah, Nanoparticle-based amelioration of drought stress and cadmium toxicity in rice via triggering the stress responsive genetic mechanisms and nutrient acquisition. Ecotoxicol. Environ. Safety **209**, 111829 (2021)

S.M. Ali, N.M.H. Yousef, N.A. Nafady, Application of biosynthesized silver nanoparticles for the control of land snail Eobaniavermiculata and some plant pathogenic fungi. Nanomater **218904**, 10 (2015)

M. Auffan, J. Rose, J.Y. Bottero, G.V. Lowry, J.P. Jolivet, M.R. Wiesner, Towards a definition of inorganic nanoparticles from an environmental, health and safety perspective. Nat. Nanotechnol. **4**, 634–664 (2009)

P. Babula, Uncommon heavy metals, metalloids and their plant toxicity: a review, in *Pest Control and Remediation of Soil Pollutants*. ed. by O. Farming (Berlin, Springer-Verlag, GmbH, 2009), pp. 275–317

S.N. Baker, G.A. Baker, Luminescent carbon nano dots: emergent nano lights. Angew. Chem. Int. Ed. **49**, 6726–6744 (2010)

P. Bansal, J.S. Duhan, S.K. Gahlawat, Biogenesis of nanoparticles: a review. Afr. J. Biotechnol. **13**, 2778–2785 (2014)

F.J. Bellesi, A.F. Arata, M. Martínez, A.C. Arrigoni, S.A. Stenglein, M.I. Dinolfo, Degradation of gluten proteins by Fusarium species and their impact on the grain quality of bread wheat. J. Stored Prod. Res. **83**, 1–8 (2019)

R. Bharati, S. Suresh, A review on nano-catalyst from waste for production of bio fuel-via-bioenergy, in, *Biofuels and Bioenergy* (Springer, Cham, 2017), pp. 25–32

D.A. Brock, T.E. Douglas, D.C. Queller, J.E. Strassmann, Primitive agriculture in asocial amoeba. Nature **469**, 393–396 (2011)

L. Bromham, C.H. Saslis-Lagoudakis, T.H. Bennett, T.J. Flowers, Soil alkalinity and salt tolerance: adapting to multiple stresses. Biol. Lett. **9**, 20130642 (2013)

M. Capuana, Heavy metal sand woody plants biotechnologies for phyto remediation. J. Biogeo. Sci. for. **4**, 7–15 (2011)

P. Chamoli, M.K. Das, K.K. Kar, Structural, optical and electrical characteristics of graphene nanosheets synthesized from microwave-assisted exfoliated graphite. J. Appl. Phys. **122**, 185105 (2017)

P. Chamoli, R.K. Shukla, A. Bezbaruah, K.K. Kar, K.K. Raina, Microwave-assisted rapid synthesis of honeycomb core-ZnO tetrapods nanocomposites for excellent photocatalytic activity against different organic dyes. Appl. Surf. Sci. **555**, 149663 (2021a)

P. Chamoli, R.K. Shukla, A. Bezbaruah, K.K. Kar, K.K. Raina, Rapid Microwave growth of mesoporous TiO_2 nano tripods for excellent photocatalysis and adsorption. J. Appl. Phys. **130**, 164901 (2021b)

N. Chartuprayoon, Y. Rheem, J.C.K. Ng, J. Nam, W. Chen, N.V. Myung, Polypyrrole nano ribbon based chemiresistive immune sensors for viral plant pathogen detection. Anal. Methods. **5**, 3497–3502 (2013)

D. Charych, Q. Cheng, A. Reichert, G. Kuziemko, N. Stroh, J. Nagy, W. Spevak, R. Stevens, A 'litmus test' for molecular recognition using artificial membranes. Chem. Biol. **3**, 113 (1996)

F. Chen, X. Hu, Study on red fermented rice with high concentration of monacolin K and low concentration of citrinin. Internat. J. Food Microbiol. **103**, 331–337 (2005)
H. Chhipa, Nanofertilizers and nanopesticides for agriculture. Environ. Chem. Lett. **15**(1), 15–22 (2017)
K.H. Cho, J.E. Park, T.O. Saka, S.G. Park, The study of antimicrobial activity and preservateive effects of nano silver ingredient. Electrochem. Acta. **51**, 956–960 (2005)
R.N. Chopra, R. Badhwar, S. Ghosh, *Poisonous Plants of India* (Indian Council of Agricultural Research, New Delhi, India, 1994)
M.K. Cik, D. Ernst, M. Komar, M. Urík, M. Sebesta, E.D. Cka, I. Cern, Y.R. Illa, R. Kanike, Y. Qian, H. Feng, D. Orlová, G. Kratosova, Effect of foliar spray application of zinc oxide nanoparticles on quantitative, nutritional, and physiological parameters of foxtail millet (Setariaitalica L.) under field conditions. Nanomaterials **9**, 1559 (2019)
M.-C. Daniel, D. Astruc, Gold nanoparticles: assembly, supramolecular chemistry, quantum-size-related properties, and applications toward biology, catalysis, and nanotechnology. Chem. Rev. **104**, 293–346 (2004)
E.H. Dehkourdi, M. Mosavi, Effect of anatase nanoparticles (TiO2) on parsley seed germination (petroselinum crispum) In Vitro. Biol. Trace Elem. Res. **155**, 283–286 (2013)
D.G. Devi, K. Murugan, P.C. Selvam, Green synthesis of silver nanoparticles using Euphorbia hirta (Euphorbiaceae) leaf extract against crop pest of cotton bollworm, Helicoverpaarmegera (Lepidoptera: Noctuidae). J. Biopest **7**, 54–66 (2014)
Ditta, How helpful is nanotechnology in agriculture? Adv. Nat. Sci. Nanosci. Nanotechnol. **3**(3), 033002 (2012)
W. Elmer, J.C. White, The future of nanotechnology in plant pathology. Annu. Rev. Phytopathol. **56**, 111–133 (2018)
K. Elumalai, S. Velmurugan, S. Ravi, V. Kathiravan, S. Ashok kumar, Green synthesis of zinc oxide nanoparticles using Moringaoleifera leaf extract and evaluation of its antimicrobial activity. Spectrochim. ActaMol. Biomol. Spectro sc. 158–164 (2015)
FAO, High level expert forum-how to feed the world in 2050. Economic and social development, food and agricultural Organization of the United Nations, Rome, Italy (2009)
C. Feigl, S. Russo, A. Barnard, Safe, stable and effective nanotechnology: phase mapping of ZnS nano-particles. J. Mater. Chem. **20**(24), 4971–4980 (2010)
W.J. Gangloff, D.G. Westfall, G.A. Peterson, J.J. Mortvedt, Mobility of organic and inorganic zinc fertilizers in soils. Commun. Soilsci. Plant Anal. **37**, 199–209 (2006)
F.Q Gao, F.S. Hong, C. Liu, L. Zheng, M.Y. Su, X. Wu, Mechanism of nanoanatase TiO2 on promoting photosynthetic carbon reaction of spinach: inducing complex of Rubisco-Rubiscoactivase. Biol. Trace Elem. Res. **11**, 239–254 (2006a)
F.Q Gao, F.S. Hong, C. Liu, L. Zheng, M.Y. Su, X. Wu, Mechanism of nanoanataseTiO2on promoting photosynthetic carbon reaction of spinach: inducing complex of Rubisco-Rubisco activase. Biol. Trace Elem. Res. **11**, 239–254 (2006b)
Y. Gao, Q. Huang, Q. Su, R. Liu, Green synthesis of silver nanoparticles at room temperature using Kiwifruit juice. Spectrosc. Lett. **47**(10), 790–795 (2014)
G. Gholamreza, A. Mohammadi, A. Aakbri, S. Panahirad, R.D. Mohammad, K. Sesuke, Titanium dioxide nanoparticles (TiO_2 NPs) promote growth and ameliorate salinity stress effects on essential oil profile and biochemical attributes of *Dracocephalum moldavica*. Sci. Rep. **10**(1), 912 (2020)
B. Gunjan, M.G.H. Zaidi, A. Sandeep, Impact of gold nano-particles on physiological and biochemical characteristics of Brassicajuncea. J. Plant Biochem. Physiol. **2**, 133 (2014)
M. Haghighi, Z. Afifipour, M. Mozafarian, The effect of N–Si on tomato seed germination under salinity levels. Intern. Environ. Sci. **6**, 87–90 (2012a)
M. Haghighi, Z. Afifipour, M. Mozafarian, The effect of N–Sion tomato seed germination unders alinity levels. Intern. Environ. Sci. **6**, 87–90 (2012b)
T. Haruyama, Micro- and nanobiotechnology for biosensing cellular responses. Adv. Drug Delivery Rev. **55**, 393–401 (2003)

H. Hasanpour, R. Maali-Amiri, H. Zeinali, Effect of TiO2 nanoparticles on metabolic limitations to photosynthesis under cold in chickpea. Russ. J. Plant Physiol. **62**, 779–787 (2015)

M. Hasanuzzaman, K. Nahar, M. Fujita, Extreme temperature responses, oxidative stress a and antioxidant defense in plants, in *Abiotic Stress Plan Responses and Applications in Agriculture*, eds. by K. Vahdati, C. Leslie (InTech Open Access Publisher, 2013)

S.S. Hojjat, Impact of silver nanoparticles on germinated fenugreek seed. Int. J. Agric. Crop. Sci. **8**, 627–630 (2015)

A.W. Hu, Z.H. Fu, Nano technology and its application in packaging and packaging machinery. Packag. Eng **24**(2005), 22–24 (2003)

S. Iravani, Green synthesis of metal nanoparticles using plants. Green Chem. **13**, 2638–2650 (2011)

C.A. Jaleel, Drought stress in plants: a review on morphological characteristics and pigments composition. Int. J. Agric. Biol. **11**, 100–105 (2009)

C. Kamaraj, G. Rajakumar, A.A. Rahuman, K. Velayutham, A. Bagavan, Feeding deterrent activity of synthesized silver nanoparticles using Manilkara zapota leaf extract against the house (2012)

S. Kantrao, M.A. Ravindra, S.M.D. Akbar, P.D.K. Jayanthi, A. Venkataraman, Effect of biosynthesized silver nanoparticles on growth and development of Helicoverpa armigera (Lepidoptera: Noctuidae): interaction with midgut protease. J. Asia Pac. Entomol. **20**(2), 583–589 (2017)

D.M. Kasote, J. Lee, G.K. Jayaprakasha, B.S. Patil, Seed priming with iron oxide nanoparticles modulate antioxidant potential and defense linked hormones in watermelon seedlings. ACS Sustain. Chem. Eng. **7**, 5142–5151 (2019)

H. Kato, In vitro assays: tracking nanoparticles inside cells. Nat. Nanotechnol. **6**, 139–140 (2011)

M. Khodakovskaya, Carbon nanotubes are able to penetrate plant seed coat and dramatically affect seed germination and plant growth. ACS Nano **3**, 3221–3227 (2009)

M.V. Khodakovskaya, de Silva, K. Biris, A.S. Dervishi, Carbon nanotubes induce growth enhancement of tobacco cells. ACS Nano **6**(3), 2128–2135 (2012)

L.R. Khot, Applications of nanomaterials in agricultural production and crop protection: a review. Crop Prot. **35**, 64–70 (2012)

Y.D. Ko, Self-supported SnO2 nano wire electrodes for high-power lithium-ion batteries. Nanotechnology **20**(45), 455701 (2009)

M.J. Lawrence, G.D. Rees, Microemulsion-based media as novel drug delivery systems. Adv. Drug Delivery Rev. **45**, 89–121 (2000)

N. Lewinski, V. Colvin, R. Drezek, CytotoXicity of nanoparticles. Small **4**, 26–49 (2008)

H.Y. Lin, C.H. Huang, S.H, Lu, I.T. Kuo, L.K. Chau, Direct detection of orchid viruses using nanorod based fiber optic particle plasma nresonance immune sensor. Biosens. Bioelectron. **51**, 371–378 (2014)

K. Logaranjan, A.J. Raiza, C.B. Subash, Y. Gopinath, Chen, shape- and size-controlled synthesis of silver nanoparticles using aloe vera plant extract and their antimicrobial activity. Nanoscale Res. Lett. **11**, 520 (2016)

T.C. Long, N. Saleh, R.D. Tilton, G.V. Lowry, B. Veronesi, Titanium dioXide (P25) produces reactive oxygen species in immortalized brain microglia (BV2): implications for nanoparticle neurotoxicity. Environ. Sci. Technol. **40**, 4346–4352 (2006)

E. López-Vargas, H. Ortega-Ortíz, G. Cadenas-Pliego, K. de Alba Romenus, M. Cabrera de la Fuente, A. Benavides-Mendoza, A. Juárez-Maldonado, Foliar application of copper nanoparticles increases the fruit quality and the content of bioactive compounds in tomatoes. Appl. Sci. **8**, 1020 (2018)

J. Lovric, S.J. Cho, F.M. Winnik, D. Maysinger, Unmodified cadmium telluride quantum dots induce reactive oxygen species formation leading to multiple organ elledamage and cell death. Chem. Biol. **12**, 1227–1234 (2005)

H.M. Siddiqui, H.M. Al-Whaibi, M. Firoz, M.Y. Al-Khaishany, Role of nanoparticles in plants. Nanotechnol. Plant Sci. 19–35 (2015)

M. Meenu, U. Kamboj, A. Sharma, P. Guha, S. Mishra, Green method for determination of phenolic compounds in mung bean (*Vigna radiata* L.) based on near-infrared spectroscopy and chemometrics. Int. J. Food Sci. Tech. **51**, 2520–2527 (2016)

S. Mishra, B.R. Singh, A. Singh, C. Keswani, A.H. Naqvi, H.B. Singh, Biofabricated silver nanoparticles act as a strong fungicide against Bipolaris sorokiniana causing spot blotch disease n wheat. PLoS ONE **9**, e97881 (2014). https://doi.org/10.1371/journal.pone.0097881

R. Mohammadi, R. MaaliAmiri, A. Abbasi, Effect of TiO2 nanoparticles on chickpea response to coldstress. Biol. Trace Elem. Res. **152**, 403–410 (2013)

J.R. Morones, J.L. Elechiguerra, A. Camacho, K. Holt, J.B. Kouri, J.T. Ramírez, M.J. Yacaman, The bactericidal effect of silver nanoparticles. Nanotechnology **16**, 2346–2354 (2005)

A. Mostofizadeh, Synthesis, properties, and applications of low-dimensional carbon-related nanomaterials. J. Nanomater. **2011**, 16 (2011)

S.S. Mukhopadhyay, Nanotechnology in agriculture: prospects and constraints. Nanotechnol. Sci. Appl. **7**, 63 (2014)

C.N. Mulligan, R.N. Yong, B.F. Gibbs, Heavy metal removal from sediments by bio surfactants. J. Hazard. Mater. **85**(1–2), 111–125 (2001)

K.R. Namburi, A.J. Kora, A. Chetukuri, V. Shree, M.K. Kota, Biogenic silver nanoparticles as an antibacterial agent against bacterial leaf, causing rice phytopathogen Xanthomonas oryzae pv. Oryzae (bioprocess. Biosyst. Eng.) **44**(9), 1975–1988 (2021). https://doi.org/10.1007/s00449-021-02579-7

M. Nuruzzaman, M.M. Rahman, Y. Liu, R. Naidu, Nanoencapsulation, nano-guard for pesticides: a new window for safe application. J. Agric. Food Chem. **64**, 1447–1483 (2016)

M. Qi, Y. Liu, T. Li, Nano TiO2 improves the photosynthesis of tomato leaves under mild heat stress. Biol. Trace Elem. Res. **156**, 323–328 (2013)

A. Patel, S. Tiwari, P. Parihar, R. Singh, S.M. Prasad, Carbon nanotubes as plant growth regulators: impacts on growth reproductive system, and soil microbial community. Nanomater. Plants, Algae Microorganisms, Concepts Controversies **2**, 23–42 (2019)

M. Patel, M. Meenu, J.K. Pandey, P. Kumar, R. Patel, Recent development in upconversion nanoparticles and their application in optogenetics: a review. J. Rare Earths (2021). https://doi.org/10.1016/j.jre.2021.10.003

K. Petersen, P.V. Nielsen, G. Bertelsen, M. Lawther, M.B. Olsen, N.H. Nilssonk, Potential of biobased materials for food packaging. Trends Food Sci. Technol. **10**, 52–68 (1999)

T.N.V.K.V. Prasad, P. Sudhakar, Y. Sreenivasulu, P. Latha, V. Munaswamy, K.R. Reddy, T.S. Sreeprasad, P.R. Sajanlal, T Pradeep, Effect of nanoscale zinc oxide particles on the germination, growth and yield of peanut. J. Plant Nutr. **35**, 905–927 (2012)

F. Puoci, F. Lemma, U.G. Spizzirri, G. Cirillo, M. Curcio, N. Picci, Polymer in agriculture: a review. Am. J. Agri. Biol. Sci. **3**, 299–314 (2008)

R. Rahimi, A. Mohammakhani, V. Roohi, N. Armand, Effects of salts tress and silicon nutrition on chlorophyll content, yield, and yield components in fennel (Foeniculum vulgare Mill). Int. J. Agric. Crop. Sci. **4**, 1591–1595 (2012)

G. Rajakumar, A.A. Rahuman, Larvicidal activity of synthesized silver nanoparticles using Ecliptaprostrata leaf extract against filarias is and malaria vectors. Acta. Trop **18**(3), 196–203 (2011)

P. Rajiv, S. Rajeshwari, R. Venckatesh, Bio-fabrication of zinc oxide nanoparticles synthesis of zinc oxide nano particles using Moringaoleifera leaf extract and evaluation of its antimicrobial activity. Spectrochim. Acta Mol. Biomol Spectrosc. **143**, 158–164 (2015)

N. Rascio, F. NavariIzzo, Heavy metal hyper accumulating plants: how and why do they do it? And what makes them so interesting? Plant Sci. 169–180 (2011)

C.M. Rico, J. Hong, M.I. Morales, L. Zhao, A.C. Zhang, J.Y. Barrios, Effect of cerium oxide nanoparticles on rice: are scence imaging. Environ. Sci. Technol. **47**, 5635 (2013)

S.W. Ryter, H.P. Kim, A. Hoetzel, J.W. Park, K. Nakahira, X. Wang, A.M. Choi, Mechanisms of cell death in oxidative stress. AntioXid. Redox Signal. **9**, 49–89 (2007)

N. Sabaghnia, M. Janmohammad, Effect of nano-silicon particles application on salinity tolerance in early growth of some lentil genotypes. Ann. UMCS. Biol. **69**(2), 39–55 (2015)

Z. Sadowski, Biosynthesis and applications of silver and gold nanoparticles, in *Silver Nanoparticles*, ed. by David Pozo Perez (2010), pp. 257–276

H. Safarpour, M.R. Safarnejad, M. Tabatabaei, A. Mohsenifar, F. Rad, M. Basirat, F. Shahryari, F. Hasanzadeh, Development of a quantum dots FRET-based biosensor for efficient detection of Polymyxabetae. Can. J. Plant Pathol. **34**, 507–515 (2012)

A. Sarkar, G. Praveen, Utilization of waste biomass into useful forms of energy, in *Biofuels and Bio-energy* (BICE (2016)) (Springer, Cham, 2017), pp. 117–132

G. Savvasd, D. Giotes, E. Chatzieustratiou, M. Bakea, G. Patakioutad, Silicon supply in soilless cultivation of Zucchini allevi-ates stress induced by salinity and powdery mildew infection. Environ. Exp. Bot. **65**, 11–17 (2009)

R. Schils, J.E. Olesen, K.C. Kersebaum, B. Rijk, M. Oberforster, V. Kalyada, M. Khitrykau, A. Gobin, H. Kirchev, V. Manolova, Cereal yield gaps across Europe. Eur. J. Agron. **101**, 109–120 (2018a)

R. Schils, J.E. Olesen, K.C. Kersebaum, B. Rijk, M. Oberforster, V. Kalyada, M. Khitrykau, A. Gobin, H. Kirchev, V. Manolova, Cereal yield gaps across Europe. Eur. J. Agron. **101**, 109–120 (2018b)

N. Shabnam, P. Pardha-Saradhi, P. Sharmila, Phenolic simpartAu3 þ-stress tolerance to cowpea by generating nanoparticles. PLoSOne **9**, 85242 (2014)

Y. Shang, K.M. Hasan, G.J. Ahammed, M. Li, H. Yin, J. Zhou, Applications of nanotechnology in plant growth and crop protection: a review. Molecules **24**, 2558 (2019)

M. Sharon, A.K. Choudhary, R. Kumar, Nanotechnology in agricultural diseases. J. Phytol. **2**, 83–85 (2010)

C. Shiva, M. Santosh, S. Kumar, Synthesized silver nanoparticles using Aristolochia indica extract against Helicoverpaarmigera Hubner (Lepidoptera:Noctuidae). Int. J. Adv. Sci. Tech. Res. **5**(2), 197–226 (2020)

T.R. Shojaei, The effect of plant growth regulators, cultivars and substrate combination on production of virus free potato mini tubers. Afr. J. Biotechnol. **8**, 4864–4871 (2009)

S. Shrivastava, D. Dash, Nanotechnology in food sector and agriculture. Proc. Natl. Acad. Sci. India Sect. BBiol. Sci. **82**(1), 29–35 (2012)

R.K. Shukla, P. Chamoli, K.K. Raina, Lyotropic liquid crystalline nano templates for synthesis of ZnS cogwheels. J. Mol. Liq. **283**, 667–673 (2019)

V. Singh, Titanium dioxide nanoparticles and its impact on growth, biomass and yield of agricultural crops under environmental stress: A review. Res. J. Nanosci. Technol. **10**, 1–8 (2020)

J. Singh, B.K. Lee, Influence of nano-TiO2 particles on the bio accumulation of Cd in soybean plants (Glycinemax): possible mechanism for the removal of Cd from the contaminated soil. J. Environ. Manag. **170**, 88–96 (2016)

S. Singh, B.K. Singh, S.M. Yadav, A.K. Gupta, Applications of nanotechnology in agricultural and their role in disease management. Res. J. Nanosci. Nanotechnol. **5**, 1–5 (2015)

P. Singh, P. Chamoli, S. Suchdev, K.K. Raina, R.K. Shukla, Structural, optical and rheological behavior investigations of graphene oxide/ glycerol based lyotropic liquid crystalline phases. Appl. Surf. Sci. **509**, 144710 (2020)

S. Sinha Ray, M. Okamoto, Polymer/layered silicate nanocomposites: a review from preparation to processing. Progress Polym. Sci. **28**, 1539–1641 (2003)

C.M. Stewart, R.B. Tompkin, M.B. Cole, Food safety: new concepts for the new millennium. Innov. Food Sci. Emerg. Technol. **3**, 105–112 (2002)

R.N. Tharanathan, Biodegradable films and composite coatings: past, present and future. Trends Food Sci. Technol. **14**(3), 71–78 (2003)

M. Thwala, N. Musee, L. Sikhwivhilu, V. Wepener, The oxidative toxicity of Ag and ZnO nanoparticles towards the aquatic plant Spirodelapunctuta and the role of testing media parameters. Environ. Sci. Process. Impacts **15**, 1830–1843 (2013)

J. Tian, K.K. Wong, C.M. Ho, C.N. Lok, W.Y. Yu, C.M. Che, J.F. Chiu, P.K. Tam, Topical delivery of silver nano particles promotes wound healing. Chem. Med. Chem. **2**, 129–136 (2007)

D.K. Tripathi, V.P. Singh, S.M. Prasad, D.K. Chauhan, N.K. Dubey, Silicon nanoparticles (SiNp) alleviate chromium(VI) phytotoxicity in Pisumsativum (L.) seedlings Plant Physiol. Biochem. **96**, 189–198 (2015)

A. Wahid, Physiological implications of metabolites biosynthesis proline assimilation and heat stress tolerance in Sugarcane (*Saccharum officinarum*) sprouts. J. Plant Res. **120**, 219–228 (2007)
X. Wang, Microwave assisted one step greensynthes is of cell-permeable multi-color photo luminescent carbon dots without surface passivation reagents. J. Mater. Chem. **21**, 2445–2450 (2011)
J. Wei, Simple one-step synthesis of water-soluble fluorescent carbon dots derived from paper ash. RSC Adv. **3**, 13119–13122 (2013)
R. Welti, W. Li, M. Li, Y. Sang, H. Biesiada, H.E. Zhou, Profiling membrane lipids in plant stress responses: role of phospholipase Dain freezing induced lipid changes in Arabidopsis. J. Biol. Chem. **277**, 31994–32002 (2002)
R.L. Whistler, J.R. Daniel, Functions of polysaccharides in foods, in *Food Additives* (Marcel Dekker, Inc., New York, NY, 1990), pp. 395–424
I.A.M. Worms, J. Boltzman, M. Garcia, V.I. Slaveykova, Cell-wall-dependent effect of carboxyl-Cd Se/Zn S quantum dots on lead and copper availability to green microalgae. Environ. Pollut. **167**, 27–33 (2012)
T. Xia, M. Kovochich, J. Brant, M. Hotze, J. Sempf, T. Oberley, C. Sioutas, J.I. Yeh, M.R. Wiesner, A.E. Nel, Comparison of the abilities of ambient and manufactured nanoparticles to induce cellular toxicity according to an oXidative stress paradigm. Nano Lett. **6**, 1794–1807 (2006)
Y. Xie, Y. He, P.L. Irwin, T. Jin, X. Shi, Antibacterial activity and mechanism of action of zinc oxide nano particles against Campylobacter jejuni. Appl. Environ. Microbiol. **77**, 2325–2331 (2011)
X. Xin, F. Zhao, J.Y. Rho, S.L. Goodrich, B.S. Sumerlin, Z. He, Use of polymeric nanoparticles to improve seed germination and plant growth under copper stress. Sci. Total Environ. **745**(25), 141055 (2020)
F. Yang, F. Hong, W. You, C. Liu, F. Gao, C. Wu, P. Yang, Influence of nano-anatase TiO2 on the nitrogen metabolism of growing spinach. Biol. Trace Elem. Res. **110**, 179–190 (2006)
X.J. Yang, X. Duan, Y. Jiang, P. Zhang, Increased expression of native cytosolic Cu/Zn superoxide dismutase and ascorbate peroxidase improves tolerance to oxidative and chilling stresses in cassava (Manihot esculenta Crantz). BMC Plant Biol. **14**, 208 (2014)
R. Yordanova, L. Popova, Effect of exogenous treatment with salicylic acid on photosynthetic activity and antioxidant capacity of chilled wheat plants. Gen. Appl. Plant Physiol. **33**, 155–170 (2007)
Y. Ze, C.L. Liu, M. Wang, F.H. Hong, The regulation ofTiO2 nanoparticles on the expression of light-harvesting complex II and photosynthesis of chloroplasts of Arabidopsis thaliana. Biol. Traceelem. Res. **143**, 1131–1141 (2011)
L. Zhao, B. Peng, J.A. Hernandez-Viezcas, C. Rico, Y. Sun, J.R. Peralta-Videa, Stress response and tolerance of Zeamaysto CeO2 nano particles: cross talk among H2O2, heat shock protein and lipid peroxidation. ACS Nano **6**, 9615–9622 (2012)
L. Zheng, F. Hong, S. Lu, C. Liu, Effect of nano-TiO (2) on strength of naturally aged seeds and growth of spinach. Biol. Trace Elem. Res. **104**, 83–91 (2005)
Zulfikar & Asraf, Nanoparticles potentially mediate salt stress tolerance in plants. Plant Physiol. Biochem. **160**, 257–268 (2021)

Prospects Toward the Development of Nanomaterials for Advanced Applications

Neha Ahlawat, Santosh Kumar Rai, and Mahesh Kumar Gupta

Abstract In this chapter, the expectations toward development in technological advancements of nanomaterials is projected through highlighting their multifunctional characteristics which enables their applications in diversified areas and key challenges in their synthesis or usage. The brief outlook about the existing applications of nanomaterials discussed in different chapters of this book provides opportunities to the researchers working toward the development of nanotechnology for explore potential of nanomaterials using multidisciplinary knowledge of science and technology. Simultaneously, efforts may also be directed to overcome existing challenges in synthesis and use of these nanomaterials in selective applications.

Keywords Nanomaterials · Opportunities · Challenges · Application areas · Perspective

Nanomaterials have become very popular in technology due to their multifunctional properties which invites their usage in diversified applications. Specially, in last two decades, the application-oriented research in nanotechnology has been growing exponentially. Their potential use in preparation of structural components, energy storage devices, sensors, biomedical and biomechanical applications, etc. has shown immense opportunities for their exploration. The bloom of nanotechnology is unexpectedly moving very fast toward medical treatments, sensor technologies, self-healing materials, big data storage and processing, weather forecasting, etc. Exposing the newer ways in medical science using wearable fitness technology using tiny

N. Ahlawat (✉)
Department of Mathematics, Jaypee Institute of Information Technology, Noida, Uttar Pradesh 201304, India
e-mail: ahlawatneha@gmail.com

S. K. Rai
Department of Mechanical Engineering, Meerut Institute of Engineering and Technology, Meerut, Uttar Pradesh 250005, India

M. K. Gupta
Department of Mechanical Engineering, SRM Institute of Science and Technology, Delhi-NCR Campus, Modinagar, Ghaziabad, Uttar Pradesh 201204, India

J. K. Katiyar et al. (eds.), *Nanomaterials for Advanced Technologies*,
https://doi.org/10.1007/978-981-19-1384-6_10

Table 1 Some popular nanomaterials and of their explored applications areas

Nanomaterial	Application areas
Gold	Medicine, food industry, water purification, etc.
Platinum	Energy storage, magnetic nanopowders cancer therapy, polymer membranes, nanocoatings, etc.
Silver	Biomedical, sensors, water/air filters, etc.
Copper	Antibiotic/antimicrobial/antifungal agent, EMI shielding, thermal conductive, conductive coating, nanometal lubricant additives, etc.
Zinc oxide	Nano-electronic/nano-optical devices, energy storage, cosmetic products, nanosensors, etc.
Carbon-based nanomaterials	Biomedical, sensors, packaging, energy storage and production, water and wastewater treatment, etc.
Titanium dioxide	Toxicity reduction in drugs, wastewater treatment, reproduction of silkworm, space applications, food industries, etc.

sensors for an intensive, accurate, and safer medical examination of living bodies is one of the important growing interests in the field of nanotechnology. Use of sensors has profound applicability in medical treatment, energy storage and harvesting, flexible manufacturing, and the development of sustainable high temperature materials. Even though the use of nanomaterials in sensing technology has introduced significant opportunities for improvement in big data analysis and whether forecasting using typical hybrid composites/nanocomposites. Table 1 comprises of some significant technological developments in diversified areas owing to the multifunctional characteristics of nanomaterials. Moreover, the potential of nanomaterials as nanofluids, nanoalloys, and self-healing materials invites researcher to explore further possibility in studying the effects of change in texture/micro-structure of the tailored materials.

Nevertheless, the expected use of nanomaterials in various high-end applications is limited due to incompetent behavior arisen by their toxic nature, compatibility issues, uniformity and repeatability in synthesis, chemical stability, and various hazardous effects to environment and living beings. Kostarelos et al. (2009) have highlighted the toxic effects of using carbon-based nanomaterials in antitumor therapies. Similarly, the effects of different nanomaterials in distortion of biological cells have been proved by various researchers (Zhang and Karn 2005; Lanone and Boczkowski 2006; Hoshino et al. 2011; Proffitt 2004; Lam et al. 2004). It has also been found that the severe effects of certain nanomaterials block the arteries in living beings and damage the central nervous system (Service RF 2004; Lam et al. 2006; Nel et al. 2006). Different aspects of challenges for realistic applications of nanotechnology are shown in Fig. 1.

The above-mentioned features and challenges in the nanomaterial's technological applicability provide an idea to explore and corelate more of such information and identify further opportunities toward the development of new/surface-modified nanomaterials or improving the existing processing methods in order to utilize their multifunctional features efficiently.

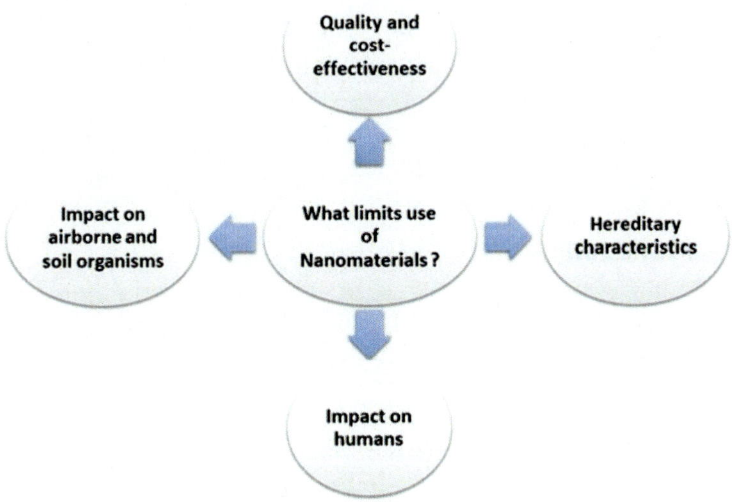

Fig. 1 Considerable challenges in application of nanomaterials for characteristic applications

References

A. Hoshino, S. Hanada, K. Yamamoto, Toxicity of nanocrystal quantum dots: the relevance of surface modifications. Arch. Toxicol. **85**(7), 707–720 (2011)

K. Kostarelos, A. Bianco, M. Prato, Promises, facts and challenges for carbon nanotubes in imaging and therapeutics. Nat. Nanotechnol. **4**(10), 627–633 (2009)

C.W. Lam, J.T. James, R. McCluskey, R.L. Hunter, Pulmonary toxicity of single-wall carbon nanotubes in mice 7 and 90 days after intratracheal instillation. Toxicol. Sci. **77**(1), 126–134 (2004)

C.W. Lam, J.T. James, R. McCluskey, S. Arepalli, R.L. Hunter, A review of carbon nanotube toxicity and assessment of potential occupational and environmental health risks. Crit. Rev. Toxicol. **36**(3), 189–217 (2006)

S. Lanone, J. Boczkowski, Biomedical applications and potential health risks of nanomaterials: molecular mechanisms. Curr. Mol. Med. **6**(6), 651–663 (2006)

A. Nel, T. Xia, L. Madler, N. Li, Toxic potential of materials at the nanolevel. Science **311**(5761), 622–627 (2006)

F. Proffitt, Yellow light for nanotech. Science **305**(5685), 762–765 (2004)

Service RF, Nanotechnology grows up. Science **304**(5678), 1732–1734 (2004)

W.X. Zhang, B. Karn, Nanoscale environmental science and technology: challenges and opportunities. Environ. Sci. Technol. **39**(5), 94A-95A (2005)